World Geography Questionnaires

Africa – Countries and Territories in the Region

Volume II

Kenneth Ma & Jennifer Fu

Geography Collections

All rights reserved. No part of this publication may be reproduced, distributed, or transmitted in any form or by any means, or stored in a database or retrieval system, without written permission.

Write to geopublish@gmail.com for more information about this book.

Book Title:	World Geography Questionnaires
	Africa – Countries and Territories in the Region
	Volume II
Author:	Kenneth Ma & Jennifer Fu
ISBN-10:	1451587074
ISBN-13:	978-1451587074

Printed in United States of America

Table of Contents

Introduction 1
Africa 2
Algeria 4
Angola 7
Benin 10
Botswana 13
Burkina Faso 17
Burundi 19
Cameroon 22
Canary Islands 25
Cape Verde 26
Central African Republic 29
Chad 31
Comoros 34
Congo, Democratic Republic of the 36
Congo, Republic of the 43
Côte d'Ivoire 46
Djibouti 48
Egypt 51
Equatorial Guinea 56
Eritrea 59
Ethiopia 61
Gabon 67
Gambia, The 70
Ghana 72
Guinea 76
Guinea-Bissau 79
Kenya 80
Lesotho 87
Liberia 89
Libya 91
Madagascar 95
Malawi 98
Mali 102
Mauritania 106
Mauritius 108
Mayotte (France) 110
Morocco 112
Mozambique 118
Namibia 121
Niger 124
Nigeria 127

- Réunion (France) 131
- Plazas de soberanía (Spain)..132
- Rwanda 134
- Saint Helena (United Kingdom) 137
- São Tomé and Príncipe 139
- Senegal...141
- Seychelles...143
- Sierra Leone 145
- Somalia 148
- South Africa...151
- Sudan 158
- Swaziland 162
- Tanzania 164
- Togo 168
- Tunisia..170
- Uganda 173
- Western Sahara 178
- Zambia 181
- Zimbabwe 184
- Miscellaneous 188
- Bibliography 197
- Other Books 197
- About the Authors 198

Introduction

As a young child, I always enjoyed looking at maps to better understand the world around me. When I heard of the National Geography Bee, I finally got a chance to put it into use. During the contest, I realized how there wasn't one study guide that covered the entire world effectively. After the contest was over, I thought that the people who had to compete after me shouldn't have to go through the same pain studying from tons of different study guides as well as helping those who simply want to learn more about the world. This got me started on this book.

This is the second of the World Geography Questionnaires series, each of which will cover questions on a different region of the world. Each book will be covering important facts about each country in that region in a question format. This book covers countries and territories in Africa – Eastern Africa, Middle Africa, Northern Africa, Southern Africa, and Western Africa.

As the second book, some improvements are made. The questions are more structured. More cross countries questions are moved to miscellaneous sections, and the country locator images are supplemented for the convenience of the reader. If you have any questions or comments, please send email to geopublish@gmail.com.

This book would have never been possible without the help of my parents, especially my mom, who is also my co-author, my teachers, and my sister. Many thanks to my parents for encouraging me and pushing me to go on, my teacher, Mr. Blair for starting and widening my interest for geography, and my sister, Hermione, who helped me find the information and get to know it better.

Kenneth Ma

Africa

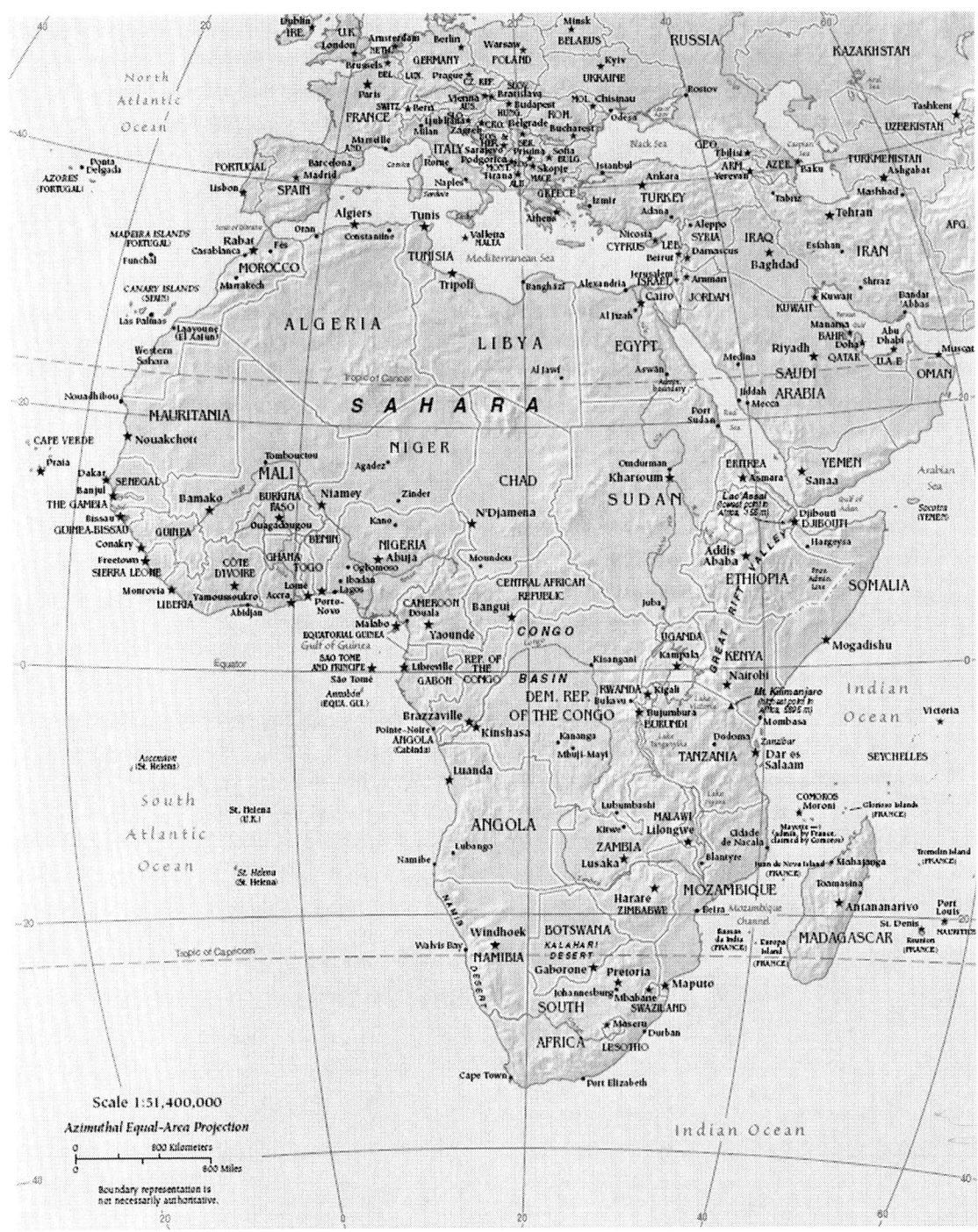

Africa is the world's second-largest and second most-populous continent, after Asia. Africa is surrounded by the Mediterranean Sea to the north, both the Suez Canal and the Red Sea along the Sinai Peninsula to the northeast, the Indian Ocean to the southeast, and the Atlantic Ocean to the west. Not counting the disputed territory of Western Sahara, there are 53 countries, including Madagascar and various island groups, associated with the continent.

The climate of Africa ranges from tropical to subarctic on its highest peaks. Its northern half is primarily desert or arid, while its central and southern areas contain both savanna plains and very dense jungle (rainforest) regions. In between, there is a convergence where vegetation patterns such as sahel, and steppe dominate.

In Africa, there are 53 countries in Africa and one territory (Western Sahara). There are also a number of small European territories.

Eastern Africa is the easterly region of the Africa. 17 countries are listed in Eastern Africa: Burundi, Comoros, Djibouti, Eritrea, Ethiopia, Kenya, Madagascar, Malawi, Mauritius, Mozambique, Rwanda, Seychelles, Somalia, Tanzania, Uganda, Zambia, and Zimbabwe. Two French overseas territories in the Indian Ocean are also included in the Eastern Africa: Réunion and Mayotte.

Middle Africa is an analogous term that describes the portion of Africa south of the Sahara Desert, east of Western Africa, but west of the Great Rift Valley. The region is dominated by the Congo River and its tributaries, which collectively drain a greater area than any river system except the Amazon. There are 9 countries in Middle Africa: Angola, Cameroon, Central African Republic, Chad, Democratic Republic of the Congo, Republic of the Congo, Equatorial Guinea, Gabon, and São Tomé and Príncipe.

Northern Africa is the northernmost region of Africa, linked by the Sahara to Sub-Saharan Africa. There are 6 countries in Northern Africa: Algeria, Egypt, Libya, Morocco, Sudan, Tunisia, and Mauritania. Western Sahara is also counted as a territory in Northern Africa. A few Spanish territories (Canary Islands, Ceuta, Melilla) and a Portuguese territory (Madeira) are also located in Northern Africa.

Southern Africa is the southernmost region of Africa. There are 5 countries in Southern Africa: Botswana, Lesotho, Namibia, South Africa, and Swaziland.

Western Africa is the westernmost region of Africa. 16 countries are listed in Western Africa: Benin, Burkina Faso, Côte d'Ivoire, Cape Verde, The Gambia, Ghana, Guinea, Guinea-Bissau, Liberia, Mali, Mauritania, Niger, Nigeria, Senegal, Sierra Leone, and Togo. Western Africa also includes the island of Saint Helena, a British overseas territory in the South Atlantic Ocean.

Algeria

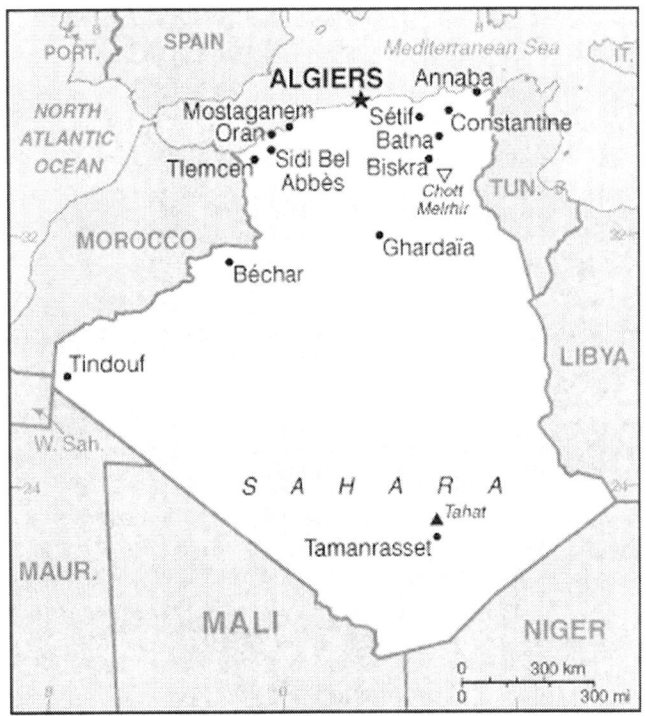
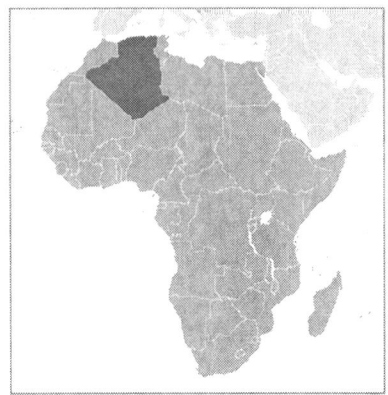

1. What is the official name of Algeria?
   People's Democratic Republic of Algeria
2. Which country borders Algeria to the west and northwest?
   Morocco
3. Which countries or territories border Algeria to the southwest?
   Mali, Mauritania, and Western Sahara
4. Which country borders Algeria to the northeast?
   Tunisia
5. Which country borders Algeria to the east?
   Libya
6. Which country borders Algeria to the southeast?
   Niger
7. Which water body borders Algeria to the north?
   The Mediterranean Sea
8. What is the motto of Algeria?
   By the people and for the people
9. What is the national anthem of Algeria?
   The Pledge
10. What is the capital of Algeria?
    Algiers (the largest city)
11. What is the official language of Algeria?

Arabic
12. What kind of government does Algeria have?
    Semi-presidential republic
13. What was Numidia, existing during 202 BCE–46 BCE?
    An ancient Berber kingdom in present-day Algeria and part of Tunisia
14. What was the capital of Numidia?
    Cirta
15. What happened to Numidia after 46 BCE?
    It alternated between being a Roman province and being a Roman client state, and it is no longer in existence today
16. During which period did Hammadid dynasty rule Algeria?
    1008–1152
17. What was the capital of Algeria?
    Beni Hammad (until 1090) and Béjaïa (after 1090)
18. Which languages were used in Hammadid dynasty?
    Classical Arabic, Berber, and Mozarabic
19. What was the religion of Hammadid dynasty?
    Sunni Islam
20. What kind of government did Hammadid dynasty have?
    Monarchy
21. Algeria was made part of which empire in 1517?
    The Ottoman Empire
22. What was the capital of Abdalwadid Dynasty during 1235–1556?
    Tlemcen
23. When did France invade Algiers?
    1830
24. When did Algeria gain independence from France?
    July 5th, 1962
25. What is the area of Algeria?
    919,595 sq mi / 2,381,741 km^2
26. What is Algeria's rank by area in Africa?
    2nd
27. How long is the coastline of Algeria?
    620 mi / 998 km
28. What is the population of Algeria?
    34,178,188 (by 2009)
29. What is the currency of Algeria?
    Algerian Dinar
30. What is the geographical feature of the terrain of Algeria?
    Mostly high plateau and desert; some mountains; narrow, discontinuous coastal plain
31. What is the highest point of Algeria?
    Mount Tahat (9,852 ft / 3,003 m)
32. Mount Tahat is in which mountain range that is a highland region in central Sahara, or southern Algeria near the Tropic of Cancer?
    The Ahaggar Mountains (also called the Hoggar)

33. What is the lowest point of Algeria?
    Chott Melrhir (-131 ft / -40 m)
34. Stretching eastward from the Moroccan border, which plateaus consist of undulating, steppe-like plains lying between the Tell and Saharan Atlas ranges?
    High Plateaus
35. The Sahara Atlas range is formed of which three massifs, sections of a planet's crust demarcated by faults or flexures?
    Ksour, Amour, and Oulad Nail
36. What is the highest point of the Aurès Mountains in eastern Algeria?
    Djebel Chélia (7,638 ft / 2,328 m)
37. The Grand Erg Oriental is a large erg in the Sahara desert. What does the erg mean?
    Field of sand dunes
38. What is the highest temperature recorded in Tiguentour, probably the highest reliable temperature ever recorded in Algeria?
    60.5 °C / 140.9 °F
39. How large is Belezma National Park, located in Batna Province of Algeria, which is famous for its 447 species of flora and 309 species of fauna?
    101 sq mi / 263 km²
40. How large is Chréa National Park, located in Blida Province of Algeria, which is home to one of the few ski resorts in Africa?
    100 sq mi / 260 km²
41. How large is Djurdjura National Park in Algeria, which is named after Djurdjura mountain chain?
    32 sq mi / 82 km²
42. How large is El Kala National Park, located in the northeast corner of Algeria, which is home to several lakes (including Lake Tonga) and a unique ecosystem in the Mediterranean basin?
    309 sq mi / 800 km²
43. How large is Gouraya National Park, located in Béjaïa Province of Algeria, which includes the 2,165 ft (660 m) high mountain of Gouraya?
    8 sq mi / 21 km²
44. Much of Tassili n'Ajjer mountain range is protected by which national park in southeastern Algeria?
    Tassili n'Ajjer National Park
45. How large is Taza National Park, located in Jijel Province of Algeria, which is famous for caves of Jijel, sand beaches, cliffs, and grottoes?
    1,470 sq mi / 3,807 km²
46. What is the highest point of Tassili n'Ajjer mountain range?
    Adrar Afao (7,080 ft / 2,158 m)
47. How large is Théniet El Had National Park, located in Tissemsilt Province of Algeria, which is home to a diversified flora and fauna?
    1,396 sq mi / 3,616 km²
48. How large is Tlemcen National Park, located in Tlemcen Province of Algeria, which includes forests, waterfalls, cliffs, and many archeoligcal sites and the ruins of Mansoura?
    32 sq mi / 82.25 km²

49. What are administrative divisions called in Algeria?
    Provinces
50. How many provinces does Algeria have?
    48
51. What is the climate of Algeria?
    Arid to semiarid; mild, wet winters with hot, dry summers along coast; drier with cold winters and hot summers on high plateau; sirocco is a hot, dust/sand-laden wind especially common in summer
52. What are the natural resources of Algeria?
    Petroleum, natural gas, iron ore, phosphates, uranium, lead, zinc
53. What are the natural hazards of Algeria?
    Mountainous areas subject to severe earthquakes; mudslides and floods in rainy season
54. What are the religions of Algeria?
    Sunni Muslim (state religion) 99%, Christian and Jewish 1%
55. What are the ethnic groups of Algeria?
    Arab-Berber 99%, European less than 1%

Angola

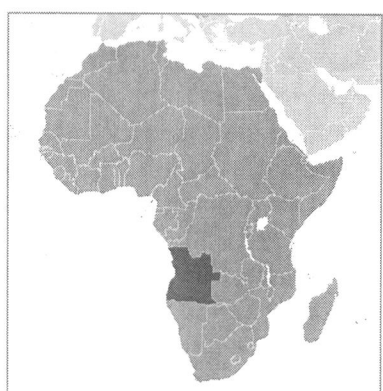

56. What is the official name of Angola?
    Republic of Angola
57. Which country borders Angola to the south?
    Namibia
58. Which country borders Angola to the north and east?

Democratic Republic of the Congo
59. Zambia borders Angola to which direction?
    East
60. Which water body borders Angola to the west?
    Atlantic Ocean
61. The exclave province of Cabinda is separated from the rest of the country by a strip, some 60 kilometres (37 mi) wide, of which country along the lower Congo River?
    Democratic Republic of Congo
62. How large is Cabinda?
    2,812 sq mi / 7,283 km^2
63. Which country borders Cabinda to the north and northeast?
    Republic of the Congo
64. Which country borders Cabinda to the east and south?
    Democratic Republic of the Congo
65. What is the motto of Angola?
    Democratic Republic of the Congo
66. What is the national anthem of Angola?
    Forward Angola!
67. What is the capital of Angola?
    Luanda (the largest city)
68. What is the official language of Angola?
    Portuguese
69. What kind of government does Angola have?
    Presidential republic
70. What names was Angola called as the Portuguese Empire's territorial expansion in South-West Africa?
    Portuguese West Africa, Portuguese Angola or the Overseas Province of Angola
71. During which period did Portuguese West Africa exist?
    1655 – 1975
72. What was the currency of Portuguese West Africa?
    Angolan Escudo
73. When did the Angolan War of Independence start?
    February 4th, 1961
74. When did Angola gain independence from Portugal?
    November 11th, 1975
75. The Angolan Civil War began after the end of the Angolan War of Independence in 1975. When did it end?
    August 2002
76. What is the area of Angola?
    481,354 sq mi / 1,246,700 km^2
77. How long is the coastline of Angola?
    994 mi / 1,600 km
78. What is the population of Angola?
    12,799,293 (by 2009)
79. What is the currency of Angola?

Kwanza
80. What is the geographical feature of the terrain of Angola?
Narrow coastal plain rises abruptly to vast interior plateau
81. What is the highest point of Angola?
Mount Moco (8,596 ft / 2,620 m)
82. What is Angola's second highest peak?
Mount Meco (8,327 ft / 2,538 m)
83. What is the lowest point of Angola?
Atlantic Ocean (0 ft / 0 m)
84. What is the longest river in Angola?
The Congo River
85. Do the Zambezi River and several tributaries of the Congo River have their sources in Angola?
Yes
86. The Cuanza River is a source of hydroelectric power of which dam?
The Capanda Dam
87. The Cunene River (also called Kunene River) flows from the Angola highlands south to the border with which country?
Namibia
88. How long is the Cunene River?
652 mi / 1,050 km
89. The Kwango River (also called Cuango) originates in the highlands of central Angola and flows generally northwards where it makes the border between Angola and which other country?
Democratic Republic of the Congo
90. How long is the Kwango River?
684 mi / 1100 km
91. The Kwango River is the tributary of which river?
The Kasai River
92. What is the largest lake in Angola?
Lago Dilolo
93. What is the largest and oldest national park in Angola?
Iona National Park (located in Namibe Province, 5,849 sq mi / 15,150 km²)
94. Iona National Park, famous for its unique flora and incredible rock formations, is contiguous with which national park in Namibia?
Skeleton Coast National Park
95. What is the smallest national park in Angola?
Cangandala National Park (243 sq mi / 630 km²)
96. Cangandala National, located in Malanje Province of Angola, was founded to protect which animal?
Giant Sable Antelope
97. How large is Bicauri National Park, located in Huila Province of Angola, which is common for riverine woodland and dry grassland is common?
305 sq mi / 790 km²

98. How large is Cameia National Park, located in Moxico Province of Angola, which has its borders with the Chifumage River to the east and the Lumege River and the Luena River to the southwest?
    558 sq mi / 1,445 km²
99. How large is Quiçama National Park (also called Kissama National Park), located in northwestern Angola, which is famous for large grasslands, tree savanna and dense thicket?
    3,846 sq mi / 9,960 km²
100. How large is Mupa National Park, located in Cunene province of Angola, which is significant for its expected wide avifauna?
    2,548 sq mi / 6,600 km²
101. What are administrative divisions called in Angola?
    Provinces
102. How many provinces does Angola have?
    18
103. What is the climate of Angola?
    Semiarid in south and along coast to Luanda; north has cool, dry season (May to October) and hot, rainy season (November to April)
104. What are the natural resources of Angola?
    Petroleum, diamonds, iron ore, phosphates, copper, feldspar, gold, bauxite, uranium
105. What are the natural hazards of Angola?
    Locally heavy rainfall causes periodic flooding on the plateau
106. What are the religions of Angola?
    Indigenous beliefs 47%, Roman Catholic 38%, Protestant 15%
107. What are the ethnic groups of Angola?
    Ovimbundu 37%, Kimbundu 25%, Bakongo 13%, Mestico 2%, European 1%, other 22%

Benin

108. What is the official name of Benin?
    Republic of Benin
109. Which country borders Benin to the west?
    Togo
110. Which country borders Benin to the east?
    Nigeria
111. Which country borders Benin to the northwest?
    Burkina Faso
112. Which country borders Benin to the north?
    Niger
113. Which water body borders Benin to the south?
    Bight of Benin
114. What is the motto of Benin?
    Fraternity, Justice, Labor
115. What is the national anthem of Benin?
    The Dawn of a New Day

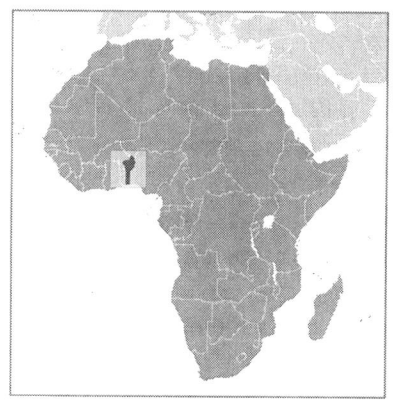

116. What is the capital of Benin?
    Porto-Novo
117. What is the largest city of Benin?
    Cotonou
118. Where is the seat of government located?
    Cotonou
119. What is the official language of Benin?
    French
120. What kind of government does Benin have?
    Multiparty democracy

11

121. Was Benin called the Benin Empire?
    No, the Benin Empire was present-day's Nigeria
122. In 1946, Benin became an overseas territory of France. What was Benin called then?
    Dahomey
123. When did Benin become Republic of Dahomey?
    December 4th, 1958
124. What was the currency of Republic of Dahomey?
    CFA Franc
125. When did Republic of Dahomey gain independence from France?
    August 1st, 1960
126. When was Republic of Dahomey renamed Benin?
    November 30th, 1975
127. Where did the name Benin come from?
    Bight of Benin
128. How was the bight named Benin?
    It was named after the Benin Empire
129. A succession of military governments ended in 1972 with the rise to power of Mathieu Kerekou and the establishment of a government based on what?
    Marxist-Leninist principles
130. A move to representative government began in 1989. Two years later, the free elections made the successful transfer of power from a dictatorship to a democracy. What was the importance?
    It was the first case in Africa that successfully transitioned from a dictatorship to a democracy
131. What is the area of Benin?
    43,484 sq mi / 112,622 km^2
132. How long is the coastline of Benin?
    75 mi / 121 km
133. Is it easy to access the coast of Benin?
    No, sandbanks create difficult access to the coast with no natural harbors, river mouths, or islands
134. What is the population of Benin?
    8,791,832 (by 2009)
135. What is the currency of Benin?
    West African CFA Franc
136. What is the geographical feature of the terrain of Benin?
    Mostly flat to undulating plain; some hills and low mountains
137. Is the low-lying, sandy, coastal plain very wide?
    No, it is at most 6 mi (10 km) wide
138. The Oueme River (also known as the Weme River) rises in which mountain?
    The Atakora Mountains
139. How long is the Oueme River, which flows past Carnotville and Oueme to a large delta on the Gulf of Guinea near the town of Cotonou?
    310 mi (500 km)

140. Are the plateaus of southern Benin split by valleys running north to south along the Couffo, Zou, and Oueme Rivers?
    Yes
141. An area of flat lands dotted with rocky hills extends around which area?
    Nikki and Save
142. What is the highest point of Benin?
    Mont Sokbaro (2,159 ft / 658 m)
143. Mont Sokbaro is located in which mountain range?
    Atacora Mountains
144. Atacora Mountains extends along the northwest border and into which country?
    Togo
145. What is the lowest point of Benin?
    Atlantic Ocean (0 ft / 0 m)
146. The Pendjari National Park lying in north western Benin is named from what?
    The Pendjari River
147. What is the Pendjari National Park known for?
    Wildlife, including elephants, monkeys, lions, hippopotamuses, buffalos, various antelopes and birds
148. What is the size of the Pendjari National Park?
    1,063 sq mi / 2,755 km^2
149. The W National Park (3,860 sq mi / 10,000 km²) is under the government of which three countries?
    Niger, Burkina Faso, and Benin
150. What are administrative divisions called in Benin?
    Departments
151. How many departments does Benin have?
    12
152. What is the climate of Benin?
    Tropical; hot, humid in south; semiarid in north
153. What are the natural resources of Benin?
    Small offshore oil deposits, limestone, marble, timber
154. What are the natural hazards of Benin?
    Hot, dry, dusty harmattan wind may affect north from December to March
155. What are the religions of Benin?
    Christian 42.8% (Catholic 27.1%, Celestial 5%, Methodist 3.2%, other Protestant 2.2%, other 5.3%), Muslim 24.4%, Vodoun 17.3%, other 15.5%
156. What are the ethnic groups of Benin?
    Fon and related 39.2%, Adja and related 15.2%, Yoruba and related 12.3%, Bariba and related 9.2%, Peulh and related 7%, Ottamari and related 6.1%, Yoa-Lokpa and related 4%, Dendi and related 2.5%, other 1.6% (includes Europeans), unspecified 2.9%

Botswana

157. What is the official name of Botswana?
    Republic of Botswana

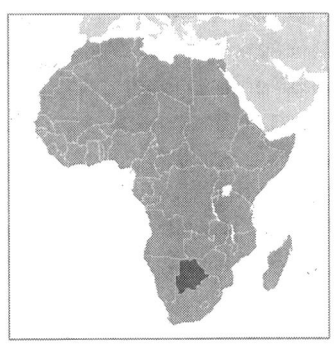

158. Which country borders Botswana to the south and southeast?
    South Africa
159. Which country borders Botswana to the west and north?
    Namibia
160. Which country borders Botswana to the northeast?
    Zimbabwe
161. How does Botswana meet Zambia?
    At a single point
162. What is the motto of Botswana?
    Rain
163. What is the national anthem of Botswana?
    This Land of Ours
164. What is the capital of Botswana?
    Gaborone (the largest city)
165. What is the official language of Botswana?
    English, Tswana
166. What kind of government does Botswana have?
    Parliamentary republic
167. When did Botswana gain independence from United Kingdom?
    September 30th, 1966
168. What is the area of Botswana?
    224,606 sq mi / 600,370 km²

169. How long is the coastline of Botswana?
    0 mi / 0 km
170. What is the population of Botswana?
    1,990,876 (by 2009)
171. What is the currency of Botswana?
    Pula
172. What is the geographical feature of the terrain of Botswana?
    Predominantly flat to gently rolling tableland; Kalahari Desert in southwest
173. What is the highest point of Botswana?
    Tsodilo Hills (4,885 ft / 1,489 m)
174. Tsodilo is a UNESCO World Heritage Site. It contains over 4,500 rock paintings in an area of approximately 4 sq miles / 10 km² within which desert?
    The Kalahari Desert
175. The Kalahari Desert covers Botswana and what other countries?
    Namibia and South Africa
176. How large is the Kalahari Desert?
    350,000 sq mi / 900,000 km²
177. The Kalahari Desert covers what percentage of Botswana?
    70%
178. Which ancient lake in present-day Botswana started evaporating many millennia ago existed in the Kalahari Desert?
    Lake Makgadikgadi
179. How large was Lake Makgadikgadi?
    80,000 km² by area and 30 m deep
180. Which rivers once emptied into Lake Makgadikgadi?
    Okavango River, Zambezi River, and Cuando River
181. What is the lowest point of Botswana?
    Junction of the Limpopo River and Shashe River (1,683 ft / 513 m)
182. The Shashe River rises from Francistown of Botswana and flows into which river?
    The Limpopo River
183. What is the world's largest inland delta?
    The Okavango Delta (also called Okavango Swamp)
184. The Okavango Delta is in which part of Botswana?
    Northwest
185. Lake Ngami is what type of lake?
    An endorheic lake (a closed drainage basin that retains water and allows no outflow to other bodies of water such as rivers or oceans)
186. Lake Ngami is one of the fragmented remnants of which ancient lake?
    Lake Makgadikgadi
187. Which river fills Lake Ngami, via the Okavango Delta?
    The Okavango River
188. Which salt pan in northern Botswana is the largest salt flat complex in the world?
    The Makgadikgadi Pan (6,178 sq mi / 16,000 km²)
189. The Makgadikgadi Pan is located which part of Botswana?
    North

190. Is the Makgadikgadi Pan a single pan?
    No, there are many pans with sandy desert in between
191. What is the area of the largest individual pan?
    1,931 sq mi / 5,000 km^2
192. What is the water source of the Makgadikgadi Pan?
    The Nata River (also called Amanzanyama in Zimbabwe)
193. The Makgadikgadi Pan formed the bed of which ancient lake?
    Lake Makgadikgadi
194. Which two national parks are located in the Makgadikgadi Pan?
    Makgadikgadi Pans National Park and Nxai Pan National Park
195. The Kgalagadi Transfrontier Park is a large wildlife preserve and conservation area in southern Africa. The park comprises which two adjoining national parks?
    Gemsbok National Park in Botswana and Kalahari Gemsbok National Park in South Africa
196. The Kgalagadi Transfrontier Park is located largely within which desert?
    Kalahari Desert
197. How large is the Kgalagadi Transfrontier Park?
    15,000 sq mi / 38,000 km^2
198. Most of Botswana's rivers cease to flow during the dry and early rainy seasons, but which rivers are exceptions?
    Chobe River, Okavango River, Boteti River and Limpopo River
199. Which national park in northwest Botswana has the world's largest concentration of African elephants?
    The Chobe National Park
200. How large is the Chobe National Park?
    4,517 sq mi / 11,700 km^2
201. What is the Moremi Game Reserve?
    A National Park, located on the eastern side of the Okavango Delta
202. How large is the Moremi Game Reserve?
    20,386 sq mi / 52,800 km^2
203. What is the second largest game reserve in the world?
    Central Kalahari Game Reserve in the Kalahari Desert of Botswana
204. What are administrative divisions called in Botswana?
    Districts and town councils
205. How many districts does Botswana have?
    9
206. How many town councils does Botswana have?
    5
207. What is the climate of Botswana?
    Semiarid; warm winters and hot summers
208. What are the natural resources of Botswana?
    Diamonds, copper, nickel, salt, soda ash, potash, coal, iron ore, silver
209. What are the natural hazards of Botswana?
    Periodic droughts; seasonal August winds blow from the west, carrying sand and dust across the country, which can obscure visibility
210. What are the religions of Botswana?

Christian 71.6%, Badimo 6%, other 1.4%, unspecified 0.4%, none 20.6%
211. What are the ethnic groups of Botswana?
    Tswana (or Setswana) 79%, Kalanga 11%, Basarwa 3%, other, including Kgalagadi and white 7%

Burkina Faso

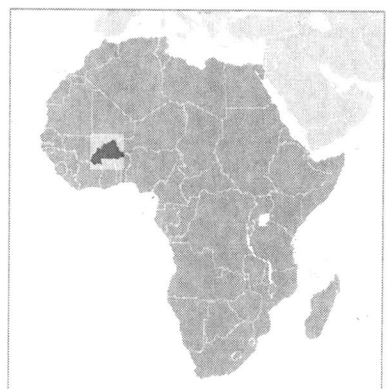

212. What is the short name of Burkina Faso?
    Burkina
213. Burkina Faso is a landlocked country, which is surrounded by how many countries?
    6
214. Which country borders Burkina Faso to the north?
    Mali
215. Which country borders Burkina Faso to the east?
    Niger
216. Which country borders Burkina Faso to the southeast?
    Benin
217. Which countries border Burkina Faso to the south?
    Togo and Ghana
218. Which country borders Burkina Faso to the southwest?
    Côte d'Ivoire
219. Which country borders Burkina Faso to the north?
    Mali
220. What is the motto of Burkina Faso?

Unity, Progress, Justice
221. What is the national anthem of Burkina Faso?
One Single Night
222. What is the capital of Burkina Faso?
Ouagadougou (the largest city)
223. What is the official language of Burkina Faso?
French
224. What kind of government does Burkina Faso have?
Semi-presidential republic
225. Upper Volta was a colony of French West Africa established on March 1st, 1919 from territories that had been part of the colonies of which countries?
Upper Senegal, Niger, and Côte d'Ivoire
226. The name Upper Volta indicates that the country contains the upper part of which river?
The Volta River
227. The Volta River passes Burkina Faso and which other country?
Ghana
228. The Volta River is divided into which three parts?
Black Volta, White Volta and Red Volta
229. Black Volta (also called Mouhoun) is a river rising in western Burkina Faso and flowing into which river?
White Volta
230. How long is Black Volta?
840 mi / 1,352 km
231. White Volta originates in Burkina Faso and flows into which lake in Ghana?
Lake Volta
232. Red Volta originates near Ouagadougou and flows about 199 mi (320 km) to join which river?
White Volta
233. Upper Volta was dissolved on September 5th, 1932 with parts being administered by which countries?
Côte d'Ivoire, French Sudan and Niger
234. When was Upper Volta revived as a part of the French Union, with its previous boundaries?
September 4th, 1947
235. When was Upper Volta reconstituted as the self-governing Republic of Upper Volta within the French Community?
December 11th, 1958
236. When did Upper Volta gain independence from France?
August 5th, 1960
237. When was Upper Volta renamed as Burkina Faso?
August 4, 1984
238. What is the area of Burkina Faso?
105,792 sq mi / 274,000 km^2
239. How long is the coastline of Burkina Faso?
0 mi / 0 km
240. What is the population of Burkina Faso?

15,746,232 (by 2009)
241. What is the currency of Burkina Faso?
West African CFA Franc
242. What is the geographical feature of the terrain of Burkina Faso?
Mostly flat to dissected, undulating plains; hills in west and southeast
243. Black Volta and which river are only two rivers in Burkina Faso flowing year-round?
Komoé River
244. Komoé River forms the border between Burkina Faso and which country?
Côte d'Ivoire
245. The basin of which river drains 27% of the country's surface?
The Niger River
246. What is the highest point of Burkina Faso?
Tena Kourou (2,457 ft / 749 m)
247. What is the lowest point of Burkina Faso?
Black Volta (656 ft / 200 m)
248. How large is Arli National Park, located in southeastern Burkina Faso, which is home to around 200 elephants, 200 hippopotamuses and 100 lions?
293 sq mi / 760 km^2
249. How large is Deux Balés National Park, located in central eastern Burkina Faso, which is an area of bushland with aging baobab trees?
313 sq mi / 810 km^2
250. Why was Pô National Park renamed to Kaboré Tambi National Park in 1991?
To honor the ranger who was killed by poachers
251. Burkina Faso and which two countries operate W National Park?
Niger and Benin
252. What are administrative divisions called in Burkina Faso?
Provinces
253. How many provinces does Burkina Faso have?
45
254. What is the climate of Burkina Faso?
Tropical; warm, dry winters; hot, wet summers
255. What are the natural resources of Burkina Faso?
Manganese, limestone, marble; small deposits of gold, phosphates, pumice, salt
256. What are the natural hazards of Burkina Faso?
Recurring droughts
257. What are the religions of Burkina Faso?
Muslim 50%, indigenous beliefs 40%, Christian (mainly Roman Catholic) 10%
258. What are the ethnic groups of Burkina Faso?
Mossi over 40%, other approximately 60% (includes Gurunsi, Senufo, Lobi, Bobo, Mande, and Fulani)

Burundi

259. What is the official name of Burundi?
Republic of Burundi

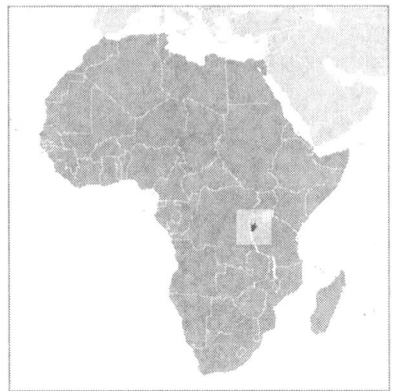

260. Which country borders Burundi to the west?
    Democratic Republic of the Congo
261. Which country borders Burundi to the north?
    Rwanda
262. Which country borders Burundi to the east and south?
    Tanzania
263. Is Burundi in the Great Lakes region?
    Yes
264. Burundi is a part of the Albertine Rift, the western extension of which valley?
    Great Rift Valley
265. The southwestern border of Burundi is adjacent to which lake?
    Lake Tanganyika
266. What is the motto of Burundi?
    Unity, Work, Progress
267. What is the national anthem of Burundi?
    Our Burundi
268. What is the capital of Burundi?
    Bujumbura (the largest city)
269. What are official languages of Burundi?
    Kirundi and French
270. What kind of government does Burundi have?
    Republic
271. Burundi was part of which colony during February 27th, 1885–June 28th, 1919?
    German East Africa
272. After World War I, Germany handed control of Burundi to which country?

Belgium
273. What was Ruanda-Urundi?
A Belgian suzerainty from 1916 to 1924, a League of Nations Class B Mandate from 1924 to 1945 and then a UN trust territory until 1962
274. What happened to Ruanda-Urundi in 1962?
It became the independent states of Rwanda and Burundi
275. When did Burundi gain independence from Belgium?
July 1st, 1962
276. Burundi's first democratically elected president was assassinated in October 1993 after how long in office?
100 days
277. Burundi's widespread ethnic violence is between which two ethnic groups?
Hutu and Tutsi
278. What is the area of Burundi?
10,745 sq mi / 27,830 km^2
279. How long is the coastline of Burundi?
0 mi / 0 km
280. What is the longest river in Burundi?
The Nile River
281. What is the population of Burundi?
8,988,091 (by 2009)
282. What is the currency of Burundi?
Burundi Franc
283. What is the geographical feature of the terrain of Burundi?
Hilly and mountainous, dropping to a plateau in east, some plains
284. What is the highest point of Burundi?
Heha (8760 ft / 2,670 m)
285. What is the lowest point of Burundi?
Lake Tanganyika (2533 ft / 772 m)
286. What are two national parks in Burundi, established in 1982?
Kibira National Park and Rurubu National Park
287. What are administrative divisions called in Burundi?
Provinces
288. How many provinces does Burundi have?
17
289. What is the climate of Burundi?
Equatorial; high plateau with considerable altitude variation (2533 ft / 772 m to 8760 ft / 2,670 m above sea level); average annual temperature varies with altitude from 73 °F / 23 °C to 63 °F / 17 °C but is generally moderate as the average altitude is about 5,577 ft / 1,700 m; average annual rainfall is about 5 ft/ 150 cm; two wet seasons (February to May and September to November), and two dry seasons (June to August and December to January)
290. What are the natural resources of Burundi?
Nickel, uranium, rare earth oxides, peat, cobalt, copper, platinum, vanadium, arable land, hydropower, niobium, tantalum, gold, tin, tungsten, kaolin, limestone
291. What are the natural hazards of Burundi?

Flooding; landslides; drought
292. What are the religions of Burundi?
Christian 67% (Roman Catholic 62%, Protestant 5%), indigenous beliefs 23%, Muslim 10%
293. What are the ethnic groups of Burundi?
Hutu (Bantu) 85%, Tutsi (Hamitic) 14%, Twa (Pygmy) 1%, Europeans 3,000, South Asians 2,000

Cameroon

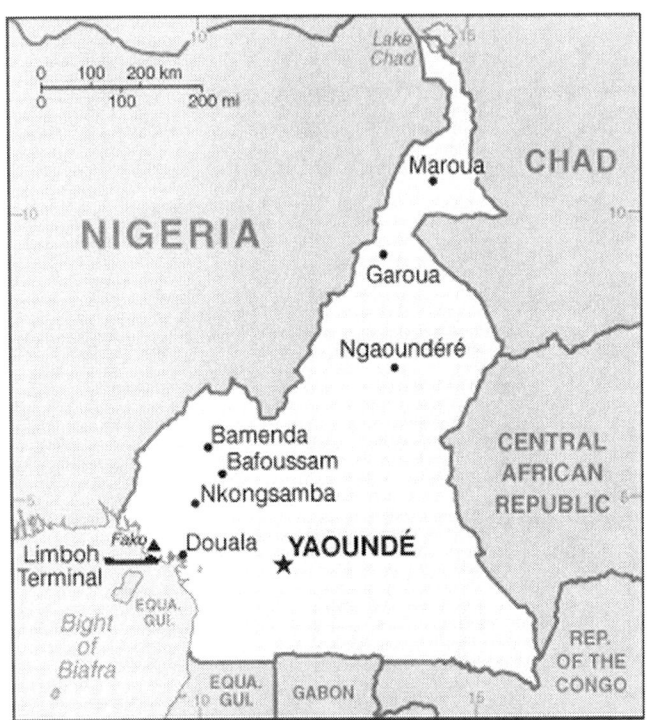
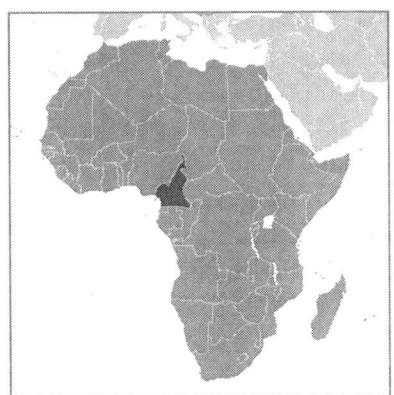

294. What is the official name of Cameroon?
Republic of Cameroon
295. Which country borders Cameroon to the west?
Nigeria
296. Which country borders Cameroon to the northeast?
Chad
297. Which country borders Cameroon to the west?
Central African Republic
298. Which countries border Cameroon to the south?
Equatorial Guinea, Gabon, and Republic of the Congo
299. Which water body borders Cameroon to the southwest?
Bight of Biafra (also called Bight of Bonny)
300. What is the motto of Cameroon?
Peace - Work - Fatherland
301. What is the national anthem of Cameroon?

O Cameroon, Cradle of our Forefathers
302. What is the capital of Cameroon?
    Yaoundé
303. What is the largest city of Cameroon?
    Douala
304. What is the official language of Cameroon?
    French, English
305. What kind of government does Cameroon have?
    Republic
306. Cameroon was integrated to which French colonial at the end of the 19th century?
    French Equatorial Africa
307. France ceded parts of the territory to German Cameroon, known as Neukamerun (Middle Congo) as a result of which crisis in 1911?
    Agadir Crisis
308. During World War I, Cameroon was occupied by troops from which countries?
    British, French and Belgian
309. Cameroon was split into which two territories in 1919?
    French Cameroun and British Cameroons
310. What was French Cameroun?
    French mandate territory in central Africa, now constituting the majority of the territory of Cameroon
311. What was British Cameroons?
    British Mandate territory in West Africa, now divided between Nigeria and Cameroon
312. What was the capital of British Cameroons?
    Buea
313. British Cameroons was administered as which two areas?
    Northern Cameroons and Southern Cameroons
314. When did French Cameroun become independent?
    January 1st, 1960
315. Northern Cameroons voted to become a region of which country on May 31st, 1961?
    Nigeria
316. Why did Southern Cameroons vote to be independent, but became a federation with Cameroon on October 1st, 1961?
    Because French Cameroun illegally occupied Southern Cameroons
317. When was Republic of Ambazonia formally declared by the Southern Cameroons Peoples Organisation (SCAPO)?
    August 31st, 2006
318. Was Republic of Ambazonia recognized by other countries or by the United Nations?
    No
319. What is the area of Cameroon?
    183,568 sq mi / 475,442 km^2
320. How long is the coastline of Cameroon?
    250 mi / 402 km
321. What is the population of Cameroon?
    18,879,301 (by 2009)

322. What is the currency of Cameroon?
    Central African CFA Franc
323. What is the geographical feature of the terrain of Cameroon?
    Diverse, with coastal plain in southwest, dissected plateau in center, mountains in west, and plains in north
324. What is the highest point of Cameroon?
    Fako (13,435 ft /4,095 m)
325. Fako is the highest peak of which mountain?
    Mount Cameroon (the highest mountain in Sub-Saharan West Africa)
326. What kind of mountain is Mount Cameroon?
    An active volcano
327. What is the Cameroon line?
    A geologic fault or rift zone that extends along the border region of eastern Nigeria and western Cameroon, from Mount Cameroon on the Gulf of Guinea north and east towards Lake Chad
328. Cameroon shares Lake Chad with which other countries?
    Chad, Nigeria, and Niger
329. Which lake is a crater lake associated with the volcanism of the Cameroon line?
    Lake Nyos
330. What is the lowest point of Cameroon?
    Atlantic Ocean (0 ft / 0 m)
331. Lake Nyos and Lake Monoun are two of the three known what type of lake?
    Exploding lake
332. Why is Cameroon called "Africa in miniature"?
    Because Cameroon exhibits all major climates and vegetation of Africa: coast, desert, mountains, rainforest, and savanna
333. Cameroon contains what national parks?
    Bénoué National Park, Bouba Njida National Park, Boumba Bek National Park, Campo Ma'an National Park, Faro National Park, Korup National Park, Lake Lobake National Park, Nki National Park, and Waza National Park
334. What are administrative divisions called in Cameroon?
    Regions
335. How many regions does Cameroon have?
    10
336. What is the climate of Cameroon?
    Varies with terrain, from tropical along coast to semiarid and hot in north
337. What are the natural resources of Cameroon?
    Petroleum, bauxite, iron ore, timber, hydropower
338. What are the natural hazards of Cameroon?
    Volcanic activity with periodic releases of poisonous gases from Lake Nyos and Lake Monoun volcanoes
339. What are the religions of Cameroon?
    Indigenous beliefs 40%, Christian 40%, Muslim 20%
340. What are the ethnic groups of Cameroon?

Cameroon Highlanders 31%, Equatorial Bantu 19%, Kirdi 11%, Fulani 10%, Northwestern Bantu 8%, Eastern Nigritic 7%, other African 13%, non-African less than 1%

Canary Islands

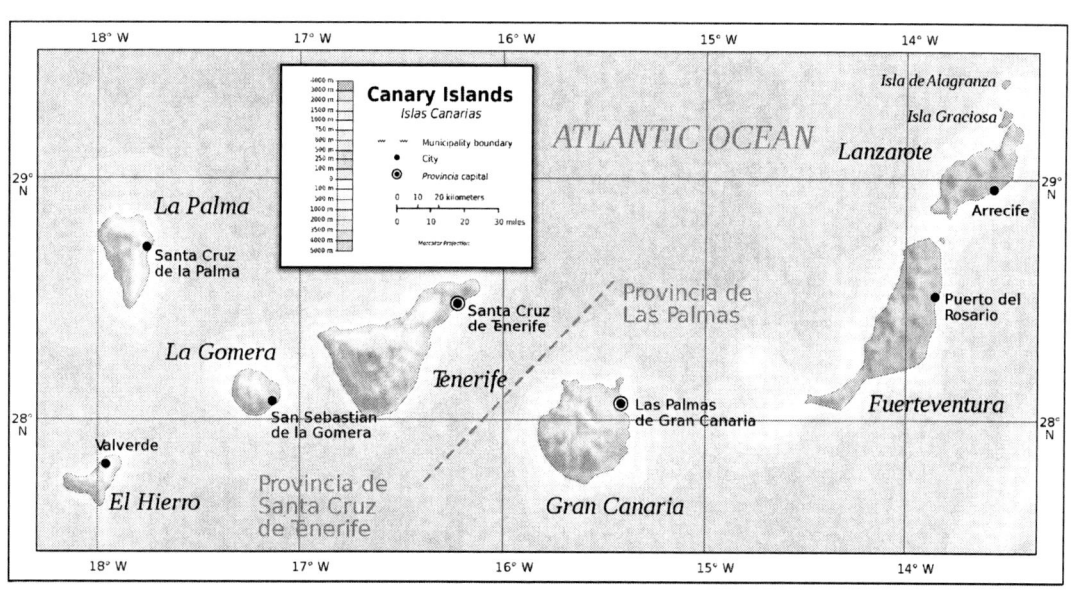

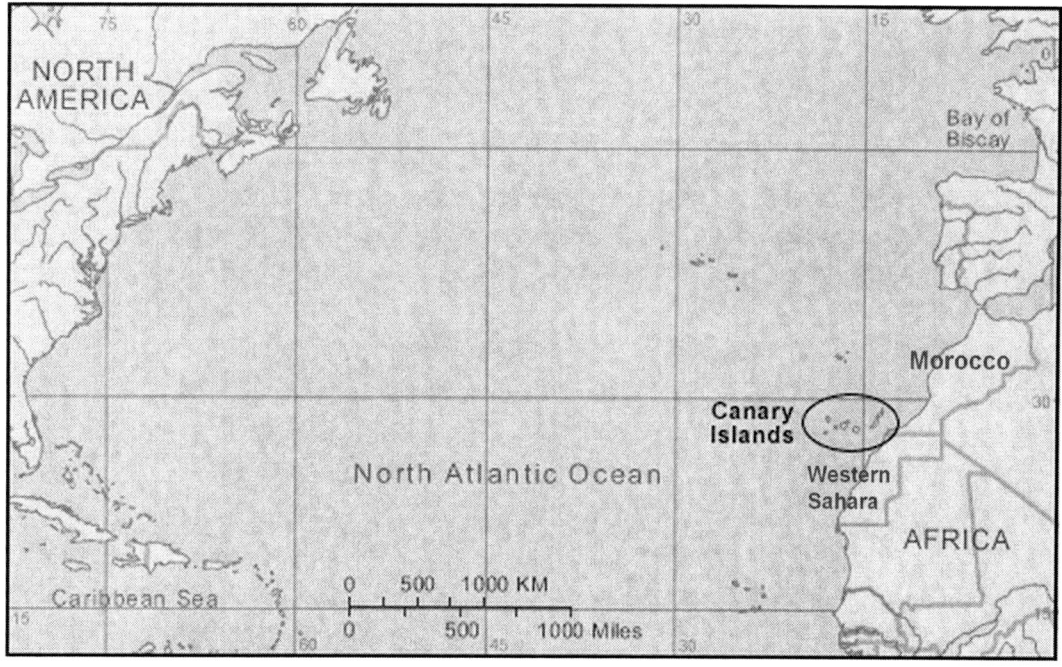

341. What are Canary Islands?
The Canary Islands are a Spanish archipelago, which forms the Spanish Autonomous Community

342. Where are Canary Islands?
    62 mi / 100 km west of the disputed border between Morocco and the Western Sahara
343. What islands are in Canary Islands?
    Tenerife, Fuerteventura, Gran Canaria, Lanzarote, La Palma, La Gomera, El Hierro, Alegranza, La Graciosa and Montaña Clara
344. What is Canary Island's most populous island, which is also Spain's most populous island?
    Tenerife (865,070 by 2009)
345. What is the national anthem of Canary Islands?
    Hymn of the Canaries
346. What is the official language of Canary Islands?
    Spanish
347. What is the area of Canary Islands?
    2,875 sq mi / 7,447 km^2
348. What is the population of Canary Islands?
    2,098,593 (by 2009)
349. What is the currency of Canary Islands?
    Euro
350. What is the geographical feature of the terrain of Canary Islands?
    Volcanic islands
351. What is the highest point of Canary Islands, which is also Spain's highest point?
    Mount Teide (12,198 ft / 3,718 m)
352. Which island is highest point of Canary Islands located?
    Tenerife
353. Mount Teide and its surroundings comprise which national park?
    The Teide National Park
354. What is the Mount Teide's rank in the world as a volcano measured by its base?
    3rd
355. Mount Teide, together with which two neighbors, form the Central Volcanic Complex?
    Pico Viejo and Montaña Blanca
356. What is the lowest point of Canary Islands?
    Atlantic Ocean (0 ft / 0 m)
357. What are administrative divisions called in Canary Islands?
    Provinces
358. What provinces do Canary Islands have?
    Las Palmas and Santa Cruz de Tenerife
359. What is the capital of Las Palmas?
    Las Palmas de Gran Canaria
360. What is the capital of Santa Cruz de Tenerife?
    Santa Cruz de Tenerife
361. What are the ethnic groups of Canary Islands?
    Spanish (Canarian and Peninsulares) 85.7%, foreign nationals 14.3%

Cape Verde

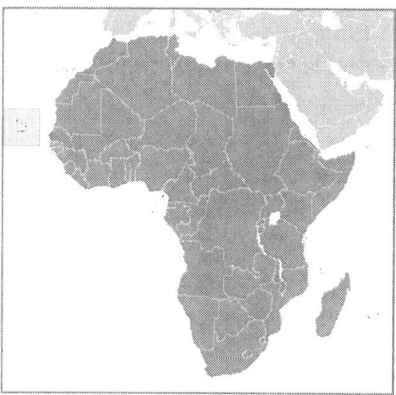

362. What is the official name of Cape Verde?
Republic of Cape Verde
363. How long is the land boundary of Cape Verde?
0 mi (0 km), it is an island country
364. Mauritania and Senegal are in which direction of Cape Verde?
East
365. How far is Cape Verde located off the west coast of Africa?
354 mi / 570 km
366. Is the westernmost point of Cape Verde also the westernmost point of Africa?
Yes
367. Cape Verde spans which ecoregion in the North Atlantic Ocean?
Macaronesia
368. What is Macaronesia?
Macaronesia consists of five archipelagos in the North Atlantic Ocean near Europe and North Africa
369. What are the five archipelagos of Macaronesia?
Azores (Portugal), Canary Islands (Spain), Cape Verde, Madeira (Portugal), Savage Islands (Portugal)
370. Cape Verde consists of how many islands, divided into the windward and leeward groups?
10 islands and 5 islets
371. How many islands are there in the windward group?
6 (Santo Antão, São Vicente, Santa Luzia, São Nicolau, Sal, and Boa Vista)
372. What are the islands in the leeward groups?
Maio, Santiago, Fogo, and Brava

373. Which island in Cape Verde is uninhabited?
    Santa Luzia
374. What is the largest island, by size and population?
    Santiago
375. What is the national anthem of Cape Verde?
    Song of Freedom
376. What is the capital of Cape Verde?
    Praia (the largest city)
377. On which island is Praia located?
    Santiago
378. What is the official language of Cape Verde?
    Portuguese
379. What kind of government does Cape Verde have?
    Republic
380. When did Portuguese settlers arrive at the uninhabited Cape Verde?
    1462
381. When did Portugal change Cape Verde's status from a colony to an overseas province in an attempt to blunt growing nationalism?
    1951
382. When did Cape Verde gain independence from Portugal?
    July 5th, 1975
383. What is the area of Cape Verde?
    1,557 sq mi / 4,033 km^2
384. How long is the coastline of Cape Verde?
    600 mi / 965 km
385. What is the population of Cape Verde?
    429,474 (by 2009)
386. What is the currency of Cape Verde?
    Cape Verdean escudo
387. What is the geographical feature of the terrain of Cape Verde?
    Steep, rugged, rocky, volcanic
388. What is the highest point of Cape Verde?
    Mount Fogo (9,282 ft / 2,829 m)
389. On which island is Mount Fogo located?
    Fogo
390. Mount Fogo is a stratovolcano mountain. Is it active?
    Yes
391. What is the lowest point of Cape Verde?
    Atlantic Ocean (0 ft / 0 m)
392. What are administrative divisions called in Cape Verde?
    Municipalities
393. How many municipalities does Cape Verde have?
    17
394. What is the climate of Cape Verde?
    Temperate; warm, dry summer; precipitation meager and very erratic

395. What are the natural resources of Cape Verde?
Salt, basalt rock, limestone, kaolin, fish, clay, gypsum
396. What are the natural hazards of Cape Verde?
Prolonged droughts; seasonal harmattan wind produces obscuring dust; volcanically and seismically active
397. What are the religions of Cape Verde?
Roman Catholic (infused with indigenous beliefs), Protestant (mostly Church of the Nazarene)
398. What are the ethnic groups of Cape Verde?
Creole (mulatto) 71%, African 28%, European 1%

Central African Republic

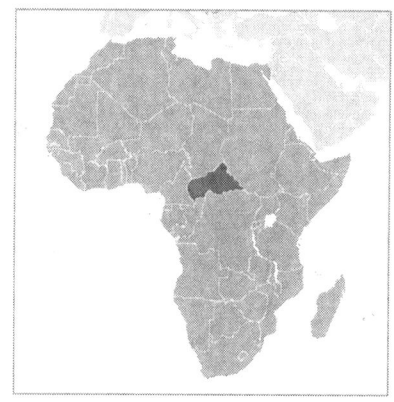

399. Which country borders Central African Republic to the west?
Cameroon
400. Which country borders Central African Republic to the east?
Sudan
401. Which country borders Central African Republic to the north?
Chad
402. Which countries border Central African Republic to the south?
Republic of the Congo and Democratic Republic of the Congo
403. Much of the southern border of Central African Republic is formed by tributaries of which river?
The Congo River

404. The Mbomou River forms part of the boundary between the Central African Republic and which other country?
Democratic Republic of the Congo
405. The Mbomou merges with the Uele River to form which river?
The Oubangui River
406. The Oubangui (also called Ubangi River) is a major tributary of which river?
The Congo River
407. Which river flows through the Central African Republic, along the border of Cameroon?
The Sangha River
408. Most of Central African Republic consists of which savannas?
Sudano-Guinean
409. Is Central African Republic almost the precise center of Africa?
Yes
410. What is the motto of Central African Republic?
Unity, Dignity, Work
411. What is the national anthem of Central African Republic?
The Renaissance
412. What is the capital of Central African Republic?
Bangui (the largest city)
413. What is the official language of Central African Republic?
Sango, French
414. What kind of government does Central African Republic have?
Republic
415. Why was Central African Republic called Oubangui-Chari (or Ubangi-Chari) while it was a French colony?
Because most of the territory is located in the Ubangi and Chari river basins
416. When did Oubangui-Chari become a semi-autonomous territory of the French Community?
1958
417. When did Central African Republic gain independence from France?
August 13th, 1960
418. What is the area of Central African Republic?
240,534 sq mi / 622,984 km^2
419. How long is the coastline of Central African Republic?
0 mi / 0 km
420. What is the population of Central African Republic?
4,511,488 (by 2009)
421. What is the currency of Central African Republic?
Central African CFA Franc
422. What is the geographical feature of the terrain of Central African Republic?
Vast, flat to rolling, monotonous plateau; scattered hills in northeast and southwest
423. What is the highest point of Central African Republic?
Mont Ngaoui (4,659 ft / 1,420 m)
424. What is the lowest point of Central African Republic?
Oubangui River (1,099 ft / 335 m)

425. How large is Manovo-Gounda St.Floris National Park, located near the Chad border, which is home to black rhinoceroses, elephants, cheetahs, leopards, red-fronted gazelles, and buffalos?
    6,718 sq mi / 17,400 km²
426. What are administrative divisions called in Central African Republic?
    Prefectures
427. How many prefectures does Central African Republic have?
    14 (plus 2 economic prefectures and 1 commune)
428. What is the climate of Central African Republic?
    Tropical; hot, dry winters; mild to hot, wet summers
429. What are the natural resources of Central African Republic?
    Diamonds, uranium, timber, gold, oil, hydropower
430. What are the natural hazards of Central African Republic?
    Hot, dry, dusty harmattan winds affect northern areas; floods are common
431. What are the religions of Central African Republic?
    Indigenous beliefs 35%, Protestant 25%, Roman Catholic 25%, Muslim 15%
432. What are the ethnic groups of Central African Republic?
    Baya 33%, Banda 27%, Mandjia 13%, Sara 10%, Mboum 7%, M'Baka 4%, Yakoma 4%, other 2%

Chad

433. What is the official name of Chad?
    Republic of Chad
434. Which country borders Chad to the north?
    Libya
435. Which country borders Chad to the east?
    Sudan
436. Which countries border Chad to the southwest?
    Cameroon and Nigeria
437. Which country borders Chad to the south?
    Central African Republic
438. Which country borders Chad to the west?
    Niger
439. Why is Chad called "Dead Heart of Africa"?
    Because of its distance from the sea and its largely desert climate
440. What is the motto of Chad?
    Unity, Work, Progress
441. What is the national anthem of Chad?
    La Tchadienne
442. What is the capital of Chad?
    N'Djamena (the largest city)
443. What is the official language of Chad?
    French, Arabic

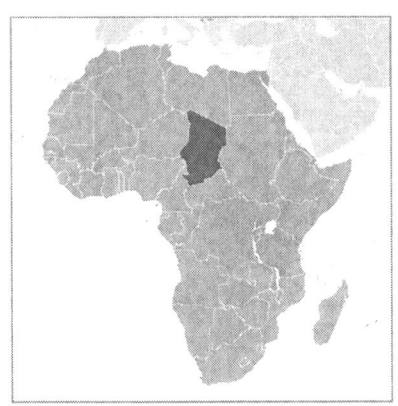

444. What kind of government does Chad have?
    Republic
445. When did Chad gain independence from France?
    August 11th, 1960
446. What is the area of Chad?
    495,753 sq mi / 1,284,000 km^2
447. How long is the coastline of Chad?
    0 mi / 0 km
448. What is the population of Chad?
    10,329,208 (by 2009)

449. What is the currency of Chad?
Central African CFA Franc
450. What is the geographical feature of the terrain of Chad?
Broad, arid plains in center, desert in north, mountains in northwest, lowlands in south
451. What is the highest point of Chad?
Emi Koussi (11,204 ft / 3,415 m)
452. What is Emi Koussi's rank by height in the Sahara?
1st
453. What type of volcano is Emi Koussi?
A dormant volcano
454. Which mountain range is Emi Koussi located?
Tibesti Mountains
455. The Tibesti Mountains are a volcanic group of inactive volcanoes with one potentially active volcano. Where are the Tibesti Mountains located?
Central Sahara desert in northern Chad
456. What is the Ennedi Plateau, located in the northeast of Chad?
A sandstone bulkwark in the middle of the Sahara
457. What is the lowest point of Chad?
Djourab Depression (525 ft / 160 m)
458. Much of Chad's population is concentrated around which river?
The Chari River (also called the Shari River)
459. How long is the Chari River that flows from the Central African Republic through Chad into Lake Chad, following the Cameroon border from N'Djamena, where it joins the Logone River waters?
590 mi / 949 km
460. What is the principle tributary of the Chari River?
The Logone River
461. What are the minor tributaries of the Chari River?
The Bahr Salamat, the Bahr Sarh, the Bahr Aouk and the Bahr Keïta
462. N'Djaména is at the spot where the Logone River empties into which river?
The Chari River
463. 90% of Lake Chad's water is fed by which river?
The Chari River
464. 10% of Lake Chad's water is fed by which river?
The Yobe River through Nigeria and Niger
465. Lake Chad is economically very important, providing water to more than 20 million people living in which four countries on the edge of the Sahara Desert?
Chad, Cameroon, Niger, and Nigeria
466. What is the surface area of Lake Chad?
521 sq mi / 1,350 km²
467. What is the surface area of Lake Chad around 4000 BCE?
154,000 sq mi / of 400,000 km² (at its largest)
468. What is the surface area of Lake Chad in 1963?
9,653 sq mi / 25,000 km²
469. Chad contains what national parks?

Aouk National Park, Goz-Beida National Park, Manda National Park, and Zakouma National Park
470. What are administrative divisions called in Chad?
    Regions
471. How many regions does Chad have?
    18
472. What is the climate of Chad?
    Tropical in south, desert in north
473. What are the natural resources of Chad?
    Petroleum, uranium, natron, kaolin, fish (Lake Chad), gold, limestone, sand and gravel, salt
474. What are the natural hazards of Chad?
    Hot, dry, dusty harmattan winds occur in north; periodic droughts; locust plagues
475. What are the religions of Chad?
    Muslim 53.1%, Catholic 20.1%, Protestant 14.2%, animist 7.3%, other 0.5%, unknown 1.7%, atheist 3.1%
476. What are the ethnic groups of Chad?
    Sara 27.7%, Arab 12.3%, Mayo-Kebbi 11.5%, Kanem-Bornou 9%, Ouaddai 8.7%, Hadjarai 6.7%, Tandjile 6.5%, Gorane 6.3%, Fitri-Batha 4.7%, other 6.4%, unknown 0.3%

Comoros

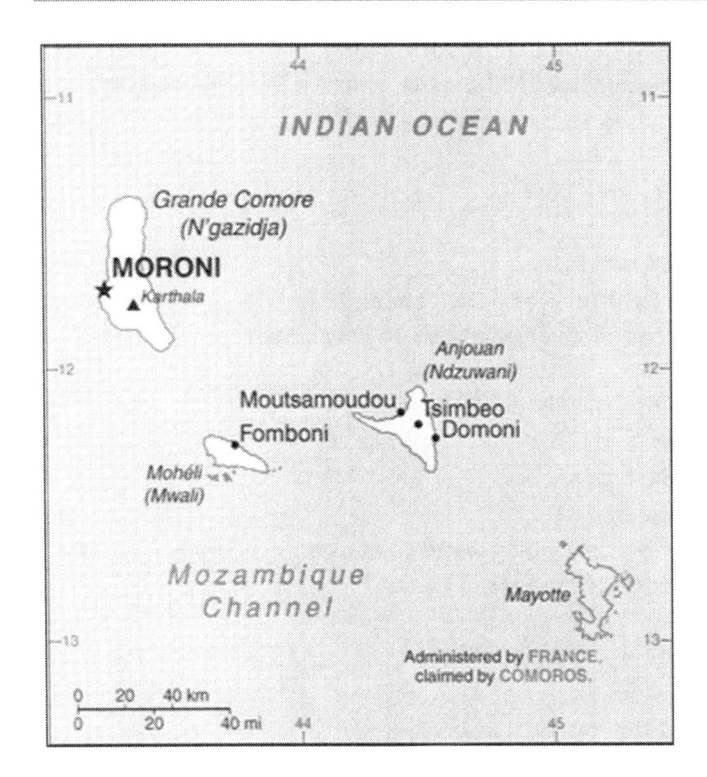

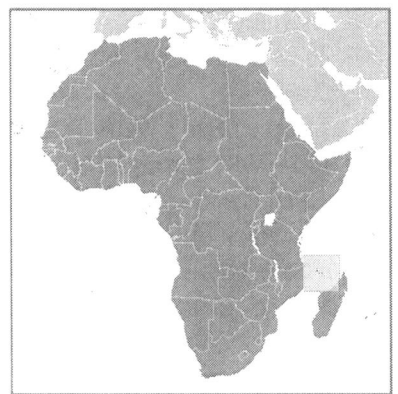

477. What is the official name of Comoros?
    Union of the Comoros
478. Comoros is an island nation in which ocean, located off the eastern coast of Africa?

Indian Ocean
479. Comoros is on the northern end of which channel, between northern Madagascar and northeastern Mozambique?
The Mozambique Channel
480. Comoros officially consists of the four main islands in the volcanic Comoros archipelago and many small islands. What are the Comorian names of the four main islands?
Ngazidja (French name: Grande Comore), Mwali (French name: Mohéli), Nzwani (French name: Anjouan), and Mahoré (French name: Mayotte)
481. Which island is the largest?
Ngazidja (442 sq mi / 1,146 km^2)
482. How large is Mwali?
112 sq mi / 290 km^2
483. How large is Nzwani?
164 sq mi / 424 km^2
484. How large is Mahoré?
145 sq mi / 375 km^2
485. What is the motto of Comoros?
Unity - Solidarity - Development
486. What is the national anthem of Comoros?
The Unity of the Islands
487. What is the capital of Comoros?
Moroni (the largest city)
488. Which island is Moroni located?
Ngazidja
489. What is the official language of Comoros?
Comorian, Arabic, French
490. What kind of government does Comoros have?
Federal republic
491. When did Comoros gain independence from France?
July 6th, 1975
492. What was Comoro's official name when declaring independence?
State of the Comoros
493. During Abdallah's presidency, what was Comoros renamed to?
Federal and Islamic Republic of Comoros
494. What is the area of Comoros?
863 sq mi / 2,235 km^2
495. How long is the coastline of Comoros?
211 mi / 340 km
496. What is the population of Comoros?
752,438 (by 2009)
497. What is the currency of Comoros?
Comorian Franc
498. What is the geographical feature of the terrain of Comoros?
Volcanic islands, interiors vary from steep mountains to low hills
499. Which two volcanoes form the Ngazidja's most prominent topographic features?

La Grille in the north and Le Kartala in the south
500. What is the highest point of Comoros?
    Le Karthala (7,743 ft / 2,360 m)
501. Is Le Karthala an active volcano?
    Yes
502. Is La Grille an active volcano?
    No
503. What is the lowest point of Comoros?
    Indian Ocean (0 ft / 0 m)
504. What kind of administrative divisions does Comoros have?
    3 islands and 4 municipalities
505. What is the climate of Comoros?
    Tropical marine; rainy season (November to May)
506. What are the natural resources of Comoros?
    Negligible
507. What are the natural hazards of Comoros?
    Cyclones possible during rainy season (December to April); Le Karthala on Grand Comore is an active volcano
508. What are the religions of Comoros?
    Sunni Muslim 98%, Roman Catholic 2%
509. What are the ethnic groups of Comoros?
    Antalote, Cafre, Makoa, Oimatsaha, Sakalava

Congo, Democratic Republic of the

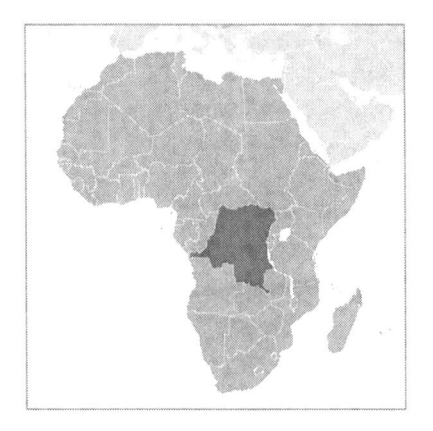

510. What are the short names of Democratic Republic of the Congo?
    DR Congo, DROC, DRC, RDC, and Congo-Kinshasa
511. Does Democratic Republic of the Congo lie on the Equator?
    Yes
512. Which countries border Democratic Republic of the Congo to the north?
    Central African Republic and Sudan
513. Which country is separated from Democratic Republic of the Congo by Lake Tanganyika to the East?
    Tanzania
514. Which countries border Democratic Republic of the Congo to the east?
    Uganda, Rwanda, and Burundi
515. Which countries border Democratic Republic of the Congo to the south?
    Zambia and Angola
516. Which country borders Democratic Republic of the Congo to the west?
    Republic of the Congo
517. Democratic Republic of the Congo has an access to which ocean through a 24.9 mi (40 km) long coastline at Muanda?
    Atlantic Ocean
518. The roughly 6 mi (9 km) wide mouth of the Congo River opens into which water body?
    Gulf of Guinea
519. Although Democratic Republic of the Congo is located in the Central African UN subregion, the nation is economically and regionally affiliated with which part of Africa?
    Southern Africa
520. What is the motto of Democratic Republic of the Congo?
    Justice – Peace – Work
521. What is the national anthem of Democratic Republic of the Congo?
    Arise Congolese
522. What is the capital of Democratic Republic of the Congo?
    Kinshasa (the largest city)
523. What was the French name of Kinshasa?
    Léopoldville
524. What is the official language of Democratic Republic of the Congo?
    French
525. What kind of government does Democratic Republic of the Congo have?
    Semi-presidential republic
526. The western portion of Democratic Republic of the Congo belonged to which kingdom during 1395–1914?
    Kingdom of Kongo
527. The southern portion of Democratic Republic of the Congo belonged to which kingdom during 1585–1889?
    Kingdom of Luba
528. What is Kingdom of Lunda during 1665–1887?
    A pre-colonial African confederation of states in what are the present-day Democratic Republic of Congo, north-eastern Angola and northwestern Zambia
529. What is Yeke Kingdom during 1856–1891?

A pre-colonial Central African state in south-east Katanga of Democratic Republic of the Congo

530. What is Kuba Kingdom during 1625–1900?
A pre-colonial Central African state bordered by Sankuru River, Lulua River, and Kasai River in the southeast of what is the present-day Democratic Republic of Congo

531. What is Colonization of the Congo during 1867–1885?
A period from Henry Morton Stanley's first exploration of the Congo until its annexation as a personal possession of the Belgium king

532. What is Congo Free State?
A corporate state privately controlled by the Belgium king through a dummy non-governmental organization

533. When did Congo Free State exist?
1885–1908

534. What was the motto of Congo Free State?
Work and Progress

535. What was the capital of Congo Free State?
Boma

536. When did Congo Free State become Belgian Congo?
November 15th, 1908

537. What was the difference between Congo Free State and Belgian Congo?
The Belgium king formally relinquished the personal control over the state to Belgium

538. When did Republic of the Congo gain independence from Belgium?
June 30th, 1960

539. What was the other name of Republic of the Congo?
Congo-Léopoldville

540. Republic of the Congo lasted until when?
1965

541. Republic of the Congo endured what crisis during June 1960 - November 1966?
The Congo Crisis

542. When did Republic of Zaire exist?
1965–1997

543. What was the national anthem of Republic of Zaire?
La Zaïroise

544. What was the currency of Republic of Zaire?
Franc until 1967, Zaïre from 1967

545. What was the result of the First Congo War from November 1996 to May 1997?
Republic of Zaire became Democratic Republic of the Congo

546. When did the Second Congo War happen?
August 2nd, 1998 – July 2003

547. Why was the Second Congo War also called Africa's World War, or Great War of Africa?
Because it was the largest war in modern African history, it directly involved eight African nations, as well as about 25 armed groups. By 2008 the war and its aftermath had killed 5.4 million people, mostly from disease and starvation, making the Second Congo War the deadliest conflict worldwide since World War II.

548. Which countries were involved in the Second Congo War?

Democratic Republic of the Congo, Namibia, Zimbabwe, Angola, and Chad vs. Uganda, Rwanda, and Burundi

549. What was the Kivu conflict?
An armed conflict between the military of the Democratic Republic of the Congo (FARDC) and the Hutu Power group Democratic Forces for the Liberation of Rwanda (FDLR) during 2004–2009

550. What is the area of Democratic Republic of the Congo?
905,355 sq mi / 2,344,858 km^2

551. What is Democratic Republic of the Congo's rank by area in Africa?
3rd

552. How long is the coastline of Democratic Republic of the Congo?
23 mi / 37 km

553. What is the population of Democratic Republic of the Congo?
68,692,542 (by 2009)

554. What is the currency of Democratic Republic of the Congo?
Congolese Franc

555. What is the geographical feature of the terrain of Democratic Republic of the Congo?
Vast central basin is a low-lying plateau; mountains in east

556. What is the highest point of Democratic Republic of the Congo?
Pic Marguerite on Mont Ngaliema (also called Mount Stanley) (16,765 ft / 5,110 m)

557. Mount Stanley is also the highest mountain of which other country?
Uganda

558. Which mountain range is Mount Stanley located?
The Rwenzori Mountains

559. What is Mount Stanley's rank by height in Africa?
3rd

560. What is the lowest point of Democratic Republic of the Congo?
Atlantic Ocean (0 ft / 0 m)

561. What are the Central Congolian lowland forests in Democratic Republic of the Congo?
An ecoregion that is a remote, inaccessible area of low-lying dense wet forest, undergrowth and swamp in the Cuvette Centrale region of the Congo Basin south of the arc of the Congo River

562. What is the Cuvette Centrale in the Democratic Republic of the Congo?
A region of forests and wetlands that is in the center of the Congo Basin, and is bounded on the west, north, and east by the arc of the Congo River

563. The Congo River and which tributaries form the backbone of Congolese economics and transportation?
Kasai River, Sangha River, Oubangui River, Aruwimi River, and Lulonga River

564. What is Lake Mweru?
A freshwater lake on the longest arm of the Congo River

565. How large is the surface area of Lake Mweru?
1977 sq mi / 5,120 km²

566. Lake Mweru is on the border between Democratic Republic of the Congo and which country?
Zambia

567. What are islands on Lake Mwer?
    Kilwa Island and Isokwe Island
568. The Lualaba River is the greatest headstream of the Congo River by which measure?
    The volume of water
569. Does the whole Lualaba River lie within Democratic Republic of the Congo?
    Yes
570. How long is the Lualaba River?
    1,118 mi / 1,800 km
571. Where does the Lualaba River become the Congo River?
    At the bottom of Boyoma Falls
572. Boyoma Falls (also called Stanley Falls) consists of how many cataracts?
    7
573. What is the total height of Boyoma Falls?
    200 ft / 60 m
574. What is the average width of Boyoma Falls?
    4,500 ft / 1,372 m
575. What is the average flow rate of Boyoma Falls?
    17,000 m³/s or 600,000 ft³/s
576. What is Pool Malebo (also called Stanley Pool, Malebo Pool, or Lake Nkunda)?
    A lake-like widening in the lower reaches of the Congo River
577. How large is the surface area of Pool Malebo?
    69 sq mi / 180 km²
578. Which two capitals are located on opposite shores of Pool Malebo?
    The capitals of the Democratic Republic of the Congo and the Republic of the Congo
579. The capitals of the Democratic Republic of the Congo and the Republic of the Congo are the two closest capital cities in the world after which other two capitals?
    Rome (Italy) and Vatican City
580. Which waterfall is the downstream from Malebo Pool?
    Livingstone Falls
581. Which three African Great Lakes lie on the Congo's eastern frontier?
    Lake Albert (also called Lake Mobutu), Lake Edward, and Lake Tanganyika
582. Lake Albert is located on the border between Democratic Republic of the Congo and which country?
    Uganda
583. What is the primary inflow of Lake Albert?
    Victoria Nile
584. What is the primary outflow of Lake Albert?
    Albert Nile
585. How large is the surface area of Lake Albert?
    2,046 sq mi / 5,300 km²
586. What is the elevation of Lake Albert?
    2,030 ft / 619 m
587. Lake Edward is located on the border between Democratic Republic of the Congo and which country?
    Uganda

588. What are the primary inflows of Lake Edward?
    Nyamugasani River, Ishasha River, Rutshuru River, and Rwindi River
589. What is the primary outflow of Lake Edward?
    Semliki River
590. How large is the surface area of Lake Edward?
    898 sq mi / 2,325 km²
591. What is the elevation of Lake Edward?
    3,018 ft / 920 m
592. Lake Kivu is one of the African Great Lakes that lies on the border between Democratic Republic of the Congo and which country?
    Rwanda
593. What is the primary outflow of Lake Kivu?
    Ruzizi River
594. How large is the surface area of Lake Kivu?
    1,040 sq mi / 2,700 km²
595. What is the elevation of Lake Kivu?
    4,790 ft / 1,460 m
596. Which island in Lake Kivu is the world's tenth-largest inland island?
    Idjwi
597. Lake Kivu is one of three known exploding lakes that undergo what rare natural disaster?
    Violent lake overturn due to volcanic activity (Other two exploding lakes are Lake Monoun and Lake Nyos)
598. What is Mount Nyiragongo of Democratic Republic of the Congo?
    A stratovolcano in the Virunga Mountains
599. Where is Mount Nyiragongo located?
    About 12 mi (20 km) north of Lake Kivu and just west of the border with Rwanda
600. What is the elevation of Mount Nyiragongo?
    11,384 ft / 470 m
601. When did Mount Nyiragongo last erupt?
    2009
602. What is Mount Nyamuragira of Democratic Republic of the Congo?
    A stratovolcano in the Virunga Mountains
603. Where is Mount Nyamuragira located?
    About 16 mi (25 km) north of Lake Kivu, 8 mi (13 km) north-north-west of Mount Nyiragongo
604. What is the elevation of Mount Nyamuragira?
    10,033 ft / 3,058 m
605. When did Mount Nyamuragira last erupt?
    2006
606. How large is Garamba National Park, home to the world's last known wild population of Northern White Rhinoceros?
    1,900 sq mi / 4,920 km²
607. How large is Kahuzi-Biéga National Park, one of the last refuges of the rare Eastern Lowland Gorilla, located in eastern Democratic Republic of the Congo?
    2,316 sq mi / 6,000 km²

608. The Kahuzi-Biéga National Park is named after which two extinct volcanoes?
Mount Kahuzi (10,853 ft / 3,308 m) and Mount Biéga (9,154 ft / 2,790 m)
609. How large is Maiko National Park, famous for its spectacular endemic animals such as the Eastern Lowland Gorilla, Okapi, and the Congo Peacock?
4,181 sq mi / 10,830 km²
610. How large is Okapi Wildlife Reserve, home to many okapis, located in the northeastern of the Democratic Republic of the Congo?
5,300 sq mi / 13,726.25 km²
611. How large is Salonga National Park, located in the Congo River basin?
13,900 sq mi / 36,000 km²
612. What is Salonga National Park's rank by area among Africa's tropical rainforest reserves?
1st
613. What animals are there in Salonga National Park?
Bonobos, Salonga monkeys, Tshuapa red colobus, Zaire peacocks, forest elephants, and African slender-snouted crocodiles
614. How large is Upemba National Park, home to some 1,800 different species, located in the Katanga province of the Democratic Republic of Congo?
4,529 sq mi / 11,730 km²
615. How large is Virunga National Park (also called Albert National Park) in the eastern Democratic Republic of Congo?
3,000 sq mi / 7,800 km²
616. What is special about Virunga National Park?
It was established in 1925 as Africa's first national park
617. What are administrative divisions called in Democratic Republic of the Congo?
Provinces
618. How many provinces does Democratic Republic of the Congo have?
10 (plus 1 city)
619. What is the climate of Democratic Republic of the Congo?
Tropical; hot and humid in equatorial river basin; cooler and drier in southern highlands; cooler and wetter in eastern highlands; north of Equator - wet season (April to October), dry season (December to February); south of Equator - wet season (November to March), dry season (April to October)
620. What are the natural resources of Democratic Republic of the Congo?
Cobalt, copper, niobium, tantalum, petroleum, industrial and gem diamonds, gold, silver, zinc, manganese, tin, uranium, coal, hydropower, timber
621. What are the natural hazards of Democratic Republic of the Congo?
Periodic droughts in south; Congo River floods (seasonal); in the east, in the Great Rift Valley, there are active volcanoes
622. What are the religions of Democratic Republic of the Congo?
Roman Catholic 50%, Protestant 20%, Kimbanguist 10%, Muslim 10%, other (includes syncretic sects and indigenous beliefs) 10%
623. What are the ethnic groups of Democratic Republic of the Congo?
Over 200 African ethnic groups of which the majority are Bantu; the four largest tribes - Mongo, Luba, Kongo (all Bantu), and the Mangbetu-Azande (Hamitic) make up about 45% of the population

Congo, Republic of the

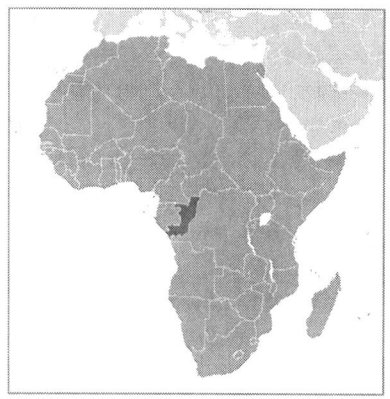

624. What are the short names of Republic of the Congo?
    Congo-Brazzaville, the Congo
625. Where is Republic of the Congo located?
    Central-western part of sub-Saharan Africa, along the Equator
626. Which country borders Republic of the Congo to the west?
    Gabon
627. Which countries border Republic of the Congo to the north?
    Cameroon and Central African Republic
628. Which country borders Republic of the Congo to the south and the east?
    Democratic Republic of the Congo
629. Which country borders Republic of the Congo to the southwest?
    Angola (exclave province of Cabinda)
630. Which water body borders Republic of the Congo to the southwest?
    Gulf of Guinea
631. What is the motto of Republic of the Congo?
    Unity, Work, Progress
632. What is the national anthem of Republic of the Congo?
    The Congolese
633. What is the capital of Republic of the Congo?
    Brazzaville (the largest city)
634. What is the official language of Republic of the Congo?

French
635. What kind of government does Republic of the Congo have?
Republic
636. Was Republic of the Congo part of Kingdom of Kongo?
Yes
637. On August 1st, 1886, France named the colony to which name?
Colony of Gabon and Congo (consisting of Gabon and Congo, which is the present-day Republic of the Congo)
638. On April 30th, 1891, France renamed the colony to which name?
Colony of French Congo (consisting of Gabon and Middle Congo, which is the present-day Republic of the Congo)
639. On January 15th, 1910, France renamed the colony to French Equatorial Africa. Which two more countries were included?
Chad and Oubangui-Chari, which is the present-day Central African Republic
640. When did Middle Congo gain autonomy from France?
November 28th, 1958
641. When did Middle Congo gain independence from France to be Republic of the Congo?
August 15th, 1960
642. People's Republic of the Congo existed during which period?
1970–1992
643. What kind of government did People's Republic of the Congo have?
Socialist republic, Single-party communist state
644. Who was the president from 1979 to 1992?
Denis Sassou Nguesso of the Congolese Labor Party
645. When was the new president, Professor Pascal Lissouba, inaugurated after the multi-party presidential election?
August 31, 1992
646. When was the Republic of the Congo Civil War?
June 1997–December 1999
647. Why did the Republic of the Congo Civil War happen?
A fight between Denis Sassou Nguesso and Pascal Lissouba
648. How many Civilians were killed during the Republic of the Congo Civil War?
Over 10,000
649. What was the result of the Republic of the Congo Civil War?
It was ended by an invasion of Angolan forces and installation of Denis Sassou Nguesso to power
650. Was Denis Sassou Nguesso still the president of Republic of the Congo at year 2009?
Yes
651. What is the area of Republic of the Congo?
132,047 sq mi / 342,000 km^2
652. How long is the coastline of Republic of the Congo?
105 mi / 169 km
653. What is the population of Republic of the Congo?
4,012,809 (by 2009)
654. What is the currency of Republic of the Congo?

Central African CFA Franc

655. What is the geographical feature of the terrain of Republic of the Congo?
Coastal plain, southern basin, central plateau, northern basin
656. What is the highest point of Republic of the Congo?
Mount Berongou (2,963 ft / 903 m)
657. What is the lowest point of Republic of the Congo?
Atlantic Ocean (0 ft / 0 m)
658. Which river forms the border between Gabon and Republic of the Congo?
The Aïna River
659. Which river forms the western-most part of the border between Democratic Republic of the Congo and Republic of the Congo?
The Chiloango River
660. Which river is the main drainage path for the coastal basin of Republic of the Congo?
The Kouilou-Niari River
661. Which river is 429 mi (600 km) long, and runs through northern Republic of the Congo?
The Nyanga River
662. Democratic Republic of the Congo and Republic of the Congo, both countries lie along which river's banks, and are named after it?
The Congo River
663. Which river flows through Central African Republic, along the border of Cameroon, and through the Republic of Congo, and empties into the Congo River?
The Sangha River
664. Which river forms the boundary between the Democratic Republic of the Congo and Republic of the Congo until it empties into the Congo River?
The Oubangui River
665. The central part of Pool Malebo is occupied by M'Bamou (also called Bamu) Island that is the territory of Republic of the Congo. How large is M'Bamou Island?
69 sq mi / 180 km²
666. What percentage of Republic of the Congo is covered by rain forest?
70%
667. How large is Nouabalé-Ndoki National Park, home to elephants and apes?
1,622 sq mi / 4,200 km²
668. How large is Odzala National Park, one of the most important strongholds for forest elephant and western gorilla conservations remaining in Central Africa?
5,251 sq mi / 13,600 Km²
669. What are administrative divisions called in Republic of the Congo?
Regions
670. How many regions does Republic of the Congo have?
10 (plus 1 commune)
671. What is the climate of Republic of the Congo?
Tropical; rainy season (March to June); dry season (June to October); persistent high temperatures and humidity; particularly enervating climate astride the Equator
672. What are the natural resources of Republic of the Congo?
Petroleum, timber, potash, lead, zinc, uranium, copper, phosphates, gold, magnesium, natural gas, hydropower

673. What are the natural hazards of Republic of the Congo?
    Seasonal flooding
674. What are the religions of Republic of the Congo?
    Christian 50%, animist 48%, Muslim 2%
675. What are the ethnic groups of Republic of the Congo?
    Kongo 48%, Sangha 20%, M'Bochi 12%, Teke 17%, Europeans and other 3%

Côte d'Ivoire

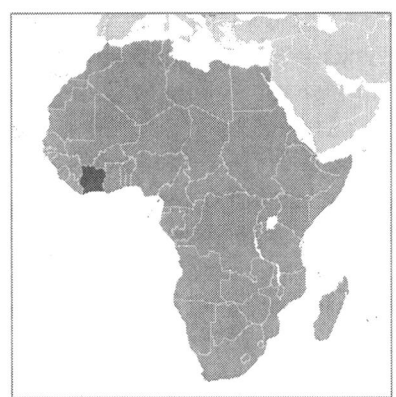

676. What is the official name of Côte d'Ivoire?
    Republic of Côte d'Ivoire
677. What does Côte d'Ivoire mean?
    Ivory Coast
678. Is Côte d'Ivoire a country of eastern sub-Saharan Africa?
    No, western sub-Saharan Africa
679. Which countries border Côte d'Ivoire to the west?
    Liberia and Guinea
680. Which countries border Côte d'Ivoire to the north?
    Mali and Burkina Faso
681. Which country borders Côte d'Ivoire to the east?
    Ghana
682. Which water body borders Côte d'Ivoire to the south?
    Gulf of Guinea
683. What is the motto of Côte d'Ivoire?

46

Unity, Discipline and Labor
684. What is the national anthem of Côte d'Ivoire?
Song of Abidjan
685. What is the capital of Côte d'Ivoire?
Yamoussoukro
686. What is the largest city of Côte d'Ivoire?
Abidjan
687. What is the official language of Côte d'Ivoire?
French
688. What kind of government does Côte d'Ivoire have?
Republic
689. Gyaman (also called Jamang) during 1450–1895 was a medieval African state of the Akan people, located in what is now Côte d'Ivoire and which country?
Ghana
690. The Kong Empire (also called Wattara Empire or Ouattara Empire) during 1710–1895 was a pre-colonial African state centered in north eastern Cote d'Ivoire that also encompassed which country?
Burkina Faso
691. When did Côte d'Ivoire officially become a French colony?
March 10th, 1893
692. Côte d'Ivoire was a constituent unit of which Federation of French colonies during 1904–1958?
Federation of French West Africa
693. When did Côte d'Ivoire gain independence from France?
August 7th, 1960
694. What is the area of Côte d'Ivoire?
124,502 sq mi / 322,460 km^2
695. How long is the coastline of Côte d'Ivoire?
320 mi / 515 km
696. What is the population of Côte d'Ivoire?
20,617,068 (by 2009)
697. What is the currency of Côte d'Ivoire?
West African CFA Franc
698. What is the geographical feature of the terrain of Côte d'Ivoire?
Mostly flat to undulating plains; mountains in northwest
699. What is the highest point of Côte d'Ivoire?
Mont Nimba (also called Mount Richard-Molard, Mount Nouon , 5,748 ft / 1,752 m)
700. Mont Nimba's peak, which is on the border of Côte d'Ivoire and another country, is the highest point of both countries. What is the other country?
Guinea
701. Mont Nimba is in the Nimba Range straddling between the border of Côte d'Ivoire with Guinea and which country?
Liberia
702. What is the lowest point of Côte d'Ivoire?
Gulf of Guinea (0 ft / 0 m)

703. What is the longest river in Côte d'Ivoire?
    The Bandama River (497 mi / 800 km)
704. Does the Bandama River flow north or south?
    South
705. Which rivers feed the Bandama River?
    Marahoué River, Solomougou River, Kan River and Nzi River
706. Where does the Bandama River empty into?
    Tagba Lagoon and the Gulf of Guinea
707. What is the largest lake in Côte d'Ivoire?
    Lake Kossou
708. Lake Kossou is an artificial lake, created in 1973 by the construction of which dam?
    Kossou Dam
709. Côte d'Ivoire contains what national parks?
    Assagny National Park, Banco National Park, Comoé National Park, Îles Ehotilés National Park, Marahoué National Park, Mont Nimba National Park, Mont Péko National Park, Mont Sângbé National Park, and Taï National Park
710. What are administrative divisions called in Côte d'Ivoire?
    Regions
711. How many regions does Côte d'Ivoire have?
    19
712. What is the climate of Côte d'Ivoire?
    Tropical along coast, semiarid in far north; three seasons - warm and dry (November to March), hot and dry (March to May), hot and wet (June to October)
713. What are the natural resources of Côte d'Ivoire?
    Petroleum, natural gas, diamonds, manganese, iron ore, cobalt, bauxite, copper, gold, nickel, tantalum, silica sand, clay, cocoa beans, coffee, palm oil, hydropower
714. What is Cote d'Ivoire's world's rank as the producer and exporter of cocoa beans?
    1st
715. What are the natural hazards of Côte d'Ivoire?
    Coast has heavy surf and no natural harbors; during the rainy season torrential flooding is possible
716. What are the religions of Côte d'Ivoire?
    Muslim 38.6%, Christian 32.8%, indigenous 11.9%, none 16.7%
717. What are the ethnic groups of Côte d'Ivoire?
    Akan 42.1%, Voltaiques or Gur 17.6%, Northern Mandes 16.5%, Krous 11%, Southern Mandes 10%, other 2.8% (includes 130,000 Lebanese and 14,000 French)

Djibouti

718. What is the official name of Djibouti?
    Republic of Djibouti
719. Is Djibouti in the Horn of Africa?
    Yes
720. Which country borders Djibouti to the west and south?
    Ethiopia

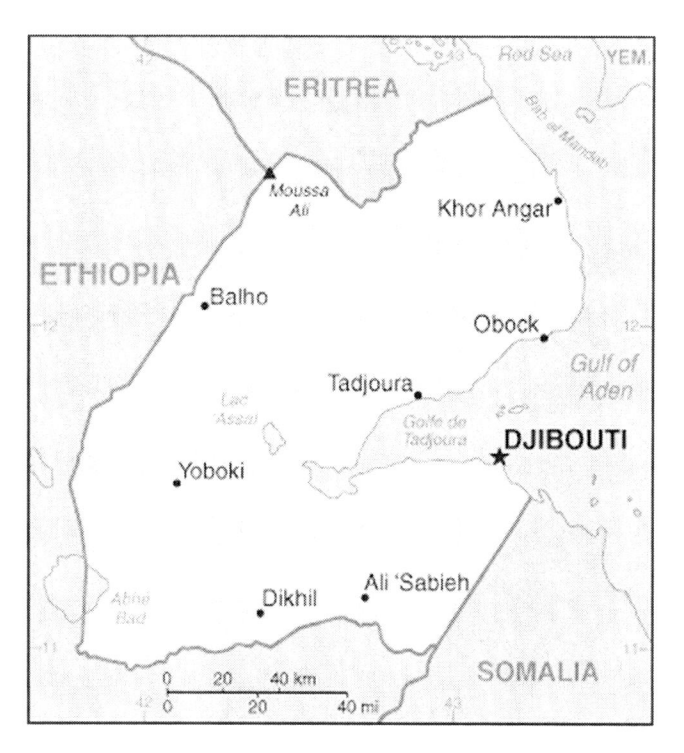

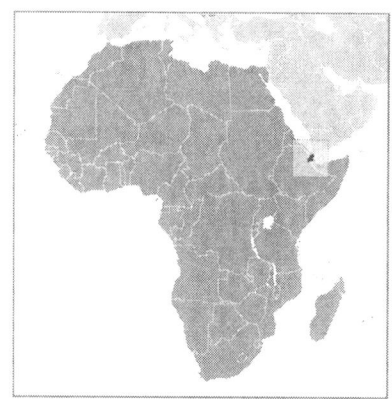

721. Which country borders Djibouti to the north?
    Eritrea
722. Which country borders Djibouti to the southeast?
    Somalia
723. Which water body borders Djibouti to the east?
    Gulf of Aden
724. What is the nickname of the Gulf of Aden due to the large amount of pirate activity in the area?
    Pirate Alley
725. Which water body borders Djibouti to the northeast?
    Red Sea
726. Which strait connects Gulf of Aden with the Red Sea?
    Bab-el-Mandeb Strait
727. How wide is the Bab-el-Mandeb Strait?
    20 mi / 32 km
728. On the other side of the Red Sea, on the Arabian Peninsula, which country is 12.4 mi (20 km) from the coast of Djibouti?
    Yemen
729. What is the motto of Djibouti?
    Unity, Equality, Peace
730. What is the national anthem of Djibouti?
    Djibouti
731. What is the capital of Djibouti?
    Djibouti (the largest city)

732. What is the official language of Djibouti?
    Arabic and French
733. What kind of government does Djibouti have?
    Semi-presidential republic
734. What was Djibouti called as a French colony during 1896—1946?
    French Somaliland
735. During which period did French Somaliland became a French Overseas territory?
    1946—1967
736. What was the name of Djibouti during 1967—1977?
    The French Territory of Afars and Issas
737. When did Djibouti gain independence from France?
    June 27th, 1977
738. What is the area of Djibouti?
    8,958 sq mi / 23,200 km^2
739. How long is the coastline of Djibouti?
    195 mi / 314 km
740. What is the population of Djibouti?
    516,055 (by 2009)
741. What is the currency of Djibouti?
    Franc
742. What is the geographical feature of the terrain of Djibouti?
    Coastal plain and plateau separated by central mountains
743. What is the highest point of Djibouti?
    Moussa Ali (6,654 ft / 2,028 m)
744. What is the lowest point of Djibouti?
    Lake Assal (-509 ft / -155 m)
745. What type of lake is Lake Assal?
    A crater lake
746. What is the lowest point of Africa?
    Lake Assal
747. What is the Lake Assal' rank among the lowest land depressions in the world?
    2nd
748. Lake Assal is considered the most saline body of water on earth outside Antarctica, with what percentage of salt concentration?
    34.8%
749. Djibouti is a part of which depression?
    Afar Depression
750. Why is Djibouti strategically important?
    Because it is located near the world's busiest shipping lanes and close to Arabian oilfields
751. Djibouti contains what national parks??
    Day Forest National Park, Djibouti National Park, and Yoboki National Park
752. What are administrative divisions called in Djibouti?
    Districts
753. How many districts does Djibouti have?
    6

754. What is the climate of Djibouti?
    Desert; torrid, dry
755. What are the natural resources of Djibouti?
    Geothermal areas, gold, clay, granite, limestone, marble, salt, diatomite, gypsum, pumice, petroleum
756. What are the natural hazards of Djibouti?
    Earthquakes; droughts; occasional cyclonic disturbances from the Indian Ocean bring heavy rains and flash floods
757. What are the religions of Djibouti?
    Muslim 94%, Christian 6%
758. What are the ethnic groups of Djibouti?
    Somali 60%, Afar 35%, other 5% (includes French, Arab, Ethiopian, and Italian)

Egypt

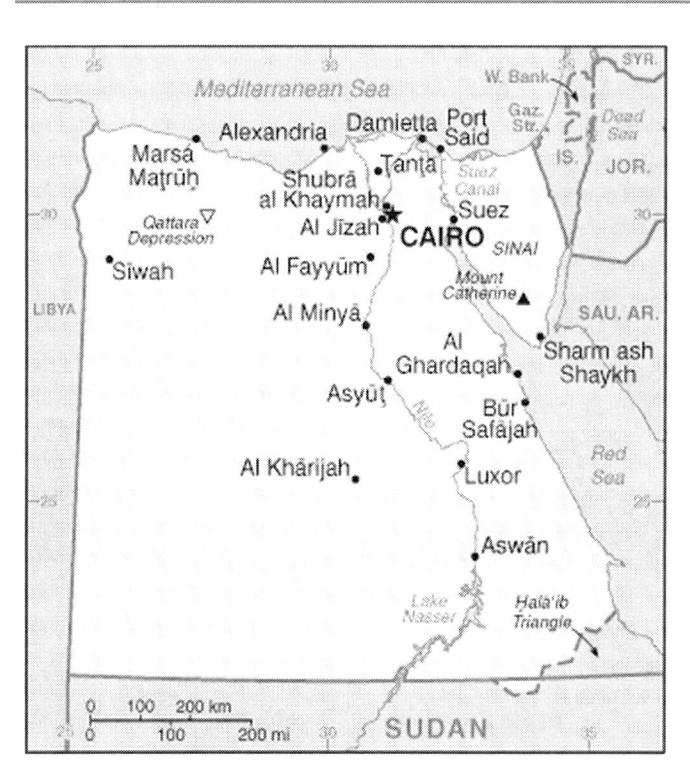

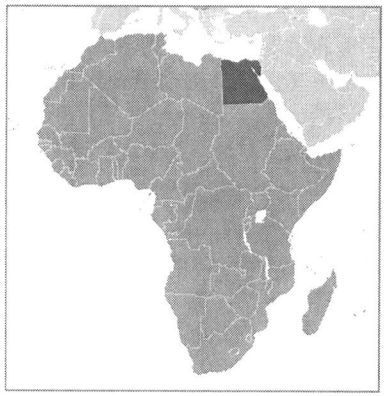

759. What is the official name of Egypt?
    Arab Republic of Egypt
760. Which country borders Egypt to the west?
    Libya
761. Which country borders Egypt to the south?
    Sudan
762. Which country and area border Egypt to the east?
    Gaza Strip and Israel
763. What is the motto of Egypt?

Life, health, well-being

764. What is the national anthem of Egypt?
My country, my country, my country
765. What is the capital of Egypt?
Cairo (the largest city)
766. What is the official language of Egypt?
Arabic
767. What kind of government does Egypt have?
Semi-presidential republic
768. Ancient Egypt was divided into which periods?
Predynastic Period (Prior to 3100 BCE), Protodynastic Period (Approximately 3100–3000 BCE), Early Dynastic Period (3100–2686 BCE), Old Kingdom (2686–2134 BCE), First Intermediate Period (2181-2055 BCE), Middle Kingdom (2080–1633 BCE), Second Intermediate Period (1640–1570 BCE), New Kingdom (1570–1070 BCE), Third Intermediate Period (also called the Libyan Period, 1070–664 BCE), Late Period (664–323 BCE)
769. The Persian Egypt existed during which three disconnected periods?
First period under Achaemenid Empire (525–402 BCE), Second period under Achaemenid Empire (336–330 BCE), and under Sassanid Empire (618–639 CE)
770. When did Ptolemaic Egypt start?
305 BCE when Ptolemy I Soter declared himself Pharaoh
771. When did Ptolemaic Egypt end?
30 BCE when Queen Cleopatra died
772. What was the capital of Ptolemaic Egypt?
Alexandria
773. What was the period after Ptolemaic Egypt?
The Roman Egypt (30 BCE–313 CE)
774. When Roman Egypt finished, which period started?
The Christian Egypt
775. What was Mamluk Sultanate during 1250–1517?
A regime composed of mamluks who ruled Egypt and Syria, and Egypt became the Muslim Egypt
776. What was the capital of Mamluk Sultanate?
Cairo
777. What was the religion of Mamluk Sultanate?
Islam
778. After Mamluk Sultanate, Egypt was under the rule of which empire?
Ottoman Empire
779. Ottoman Egypt was conquered by whom in 1798?
The French
780. Who took control of Egypt in 1805?
Albanian Muhammad Ali
781. When did the British invade and Egypt became a de facto protectorate of the United Kingdom?
1882
782. When did Egypt gain independence from United Kingdom?

February 28th, 1922

783. British troops remained in Egypt until 1956 after which treaty?
 1936 Anglo-Egyptian Treaty
784. When did Egypt become a republic?
 1952
785. What is the area of Egypt?
 387,048 sq mi / 1,002,450 km²
786. How long is the coastline of Egypt?
 1,522 mi / 2,450 km
787. What is the population of Egypt?
 83,082,869 (by 2009)
788. What is the currency of Egypt?
 Egyptian Pound
789. What is the geographical feature of the terrain of Egypt?
 Vast desert plateau interrupted by Nile Valley (also called Upper Egypt) and Delta (also called Lower Egypt)
790. What is the highest point of Egypt?
 Saint Catherine Mountain (8,625 ft / 2,629 m)
791. Due to the aridity of Egypt's climate, where is the Egyptian population concentrated?
 The Nile Valley and Delta (approximately 99% of the population uses only about 5.5% of the total land area)
792. Is the Nile Valley and Delta the most extensive oasis on earth?
 Yes
793. Which desert covers about 270,271 sq mi (700,000 km²) and accounts for about two-thirds of Egypt's land area?
 The Western Desert (also called the Libyan Desert)
794. There are eight depressions in the eastern of Western Desert; seven of them are considered oases. Why is the smallest one, Qattara, located in Egypt, not considered to be an oasis?
 Because its waters are salty
795. What is the lowest point of Egypt?
 Qattara Depression (-436 ft / -133 m)
796. Is Qattara Depression the lowest point in Africa?
 No, it is the second lowest point in Africa
797. Which plateau rises to around 3,609 ft (1,100 m) in the south of Western Desert and lies in the south west corner of Egypt?
 Gilf Kebir
798. Which Desert is east of the river Nile, between the Nile and the Red Sea?
 The Eastern Desert (also known as the Arabian Desert)
799. The Eastern Desert covers which four countries?
 Egypt, Eritrea, Sudan and Ethiopia
800. What is Sinai Peninsula?
 A triangular peninsula in Egypt, lying between the Mediterranean Sea to the north and the Red Sea to the south, forming a land bridge (the Isthmus of Suez) between Africa and Southwest Asia

801. What size is Sinai Peninsula?
    23,166 sq mi / 60,000 km²
802. Why is Egypt a transcontinental nation?
    Because it is in both Africa and Asia
803. What is the Suez Canal in Egypt?
    A man-made sea-level waterway, connecting the Mediterranean Sea and the Red Sea
804. What is the significance of the Suez Canal?
    It allows water transportation between Europe and Asia without navigating around Africa
805. When was the Suez Canal construction started?
    April 25th, 1859
806. When did the Suez Canal construction open?
    November 16th, 1869
807. Who operated and owned the Suez Canal?
    Suez Canal Company (French private investors were the majority of the shareholders, with Egypt also having a significant stake)
808. When did Britain take control of the canal by purchasing the shares of Suez Canal Company?
    August 25th, 1882
809. How long is the Suez Canal?
    119 mi / 192 km
810. What is the Great Bitter Lake?
    A salt water lake between the north and south part of the Suez Canal
811. The Great Bitter Lake is adjoined by which other lake?
    The Small Bitter Lake
812. What is the surface area of both Bitter Lakes?
    97 sq mi / 1,410 km²
813. What is the Lake Manzala?
    A brackish lake in the northern part of the Suez Canal
814. What is the surface area of the Lake Manzala?
    544 sq mi / 250 km² (very shallow)
815. What is the Lake Timsah (also called Crocodile Lake)?
    A Hydrographic lake in the southern part of the Suez Canal
816. What is the surface area of the Lake Timsah?
    5.4 sq mi / 14 km²
817. What is the Nile Delta?
    A large delta formed in Northern Egypt where the Nile River spreads out and drains into the Mediterranean Sea
818. Which Lake among three natural lakes intersected by the Suez Canal is located in the Nile Delta?
    The Lake Manzala
819. Does the Suez Canal have a lock to prevent sea water flows into the lakes from the Mediterranean and the Red Sea?
    No
820. Aswan Dam refers to which two dams, both located near Aswan in Egypt?
    High Dam and Old Aswan Dam (also called Aswan Low Dam)

821. Why did US withdraw financial aid for Aswan Dam project on July 19th, 1956?
Because of Egypt's increased ties to USSR
822. When did Egypt nationalize the Suez Canal from Suez Canal Company to the Suez Canal Authority?
July 26th, 1956
823. What was the Suez Crisis (also called Tripartite Aggression)?
A military attack on Egypt by Britain, France, and Israel caused by the nationalization of the Suez Canal
824. When did the Suez Crisis start?
October 29th, 1956
825. When did the Suez Crisis finish?
March 1957
826. What was the result of the Suez Crisis?
The Suez Canal was restored to Egypt, and Egypt agreed to pay millions of dollars to Suez Canal Company
827. During which war June 5th-10th, 1967, the Suez Canal was closed, leaving 14 ships trapped in the lake until 1975?
The Six-Day War
828. What was the Six-Day War (also called 1967 Arab-Israeli War, the Third Arab-Israeli War, Six Days' War, an-Naksah, or the June War)?
A war between Israeli and the armies of the neighboring states of 9 Arab counties (Egypt, Jordan, Syria, Iraq, Saudi Arabia, Sudan, Tunisia, Morocco and Algeria)
829. What was the result of Six-Day War?
Israel had gained control of the Sinai Peninsula, the Gaza Strip, the West Bank, East Jerusalem, and the Golan Heights
830. The northern end of the Red Sea is bifurcated by the Sinai Peninsula, creating which two gulfs?
The Gulf of Suez and the Gulf of Aqaba
831. How long is the Gulf of Suez occupying the northwestern arm of the Red Sea?
195 mi / 314 km
832. How long is the Gulf of Aqaba occupying the northeastern arm of the Red Sea?
99 mi / 160 km
833. What is the only one of the Seven Wonders of the Ancient World that survives substantially intact?
The Great Pyramid of Giza (also called the Pyramid of Khufu or the Pyramid of Cheops)
834. The Great Pyramid is the oldest and largest of the three pyramids in the Giza Necropolis. When was it built?
2,560 BCE
835. What is Giza Necropolis?
A complex of ancient monuments stands on the Giza Plateau. It includes the three pyramids known as the Great Pyramids, along with the massive sculpture known as the Great Sphinx
836. What is the Great Sphinx?
The oldest known monumental sculpture of a reclining lion with a human head
837. Why was Luxor famous?

Because it was the ancient city of Thebes, the great capital of Egypt during the New Kingdom, and the glorious city of the God Amen

838. What is the Valley of the Kings in Luxor?
A valley where, for a period of nearly 500 years from the 16th to 11th century BCE, tombs were constructed for the kings and powerful nobles of the New Kingdom

839. What is Karnak Temple Complex in Luxor?
A vast conglomeration of ruined temples, chapels, pylons and other buildings, notably the Great Temple of Amen and a massive structure begun by Pharaoh Amenhotep III

840. What is the Hala'ib Triangle (also called Sudan Government Administration Area or SGAA)?
A triangular area of land (7,950 sq mi / 20,580 km^2), located on the Red Sea's African coast, which is administered by Egypt and claimed by Sudan

841. How large is Gebel Elba National Park, which covers most of the disputed Hala'ib Triangle?
13,745 sq mi / 35,600 km^2

842. Gebel Elba is a peak in mountainous area in Hala'ib Triangle. The average annual rainfall in the region is less than 2 inches (50 mm), but what is the average annual rainfall in the upper areas of Gebel Elba?
15.7 inches (400 mm)

843. How large is Ras Muhammad National Park, located at the southern extreme of Sinai Peninsula, which is a marine reserve for the protection of marine and terrestrial wildlife?
185 sq mi / 480 km² (52 sq mi / 135 km² land area and 133 sq mi / 345 km² water area)

844. How large is Saint Catherine National Park, located in mid-southern Sinai, which is the conservation of biodiversity?
2,220 sq mi / 5,750 km^2

845. How large is Wadi El Gemal National Park, located on Red Sea coast, which is home to 240 species of coral reefs and 1,000 species of fish and marine invertebrates?
772 sq mi / 2,000 km^2

846. What are administrative divisions called in Egypt?
Governorates

847. How many governorates does Egypt have?
26

848. What is the climate of Egypt?
Desert; hot, dry summers with moderate winters

849. What are the natural resources of Egypt?
Petroleum, natural gas, iron ore, phosphates, manganese, limestone, gypsum, talc, asbestos, lead, zinc

850. What are the natural hazards of Egypt?
Periodic droughts; frequent earthquakes; flash floods; landslides; hot, driving windstorm called khamsin occurs in spring; dust storms; sandstorms

851. What are the religions of Egypt?
Muslim (mostly Sunni) 90%, Coptic 9%, other Christian 1%

852. What are the ethnic groups of Egypt?
Egyptian 99.6%, other 0.4%

Equatorial Guinea

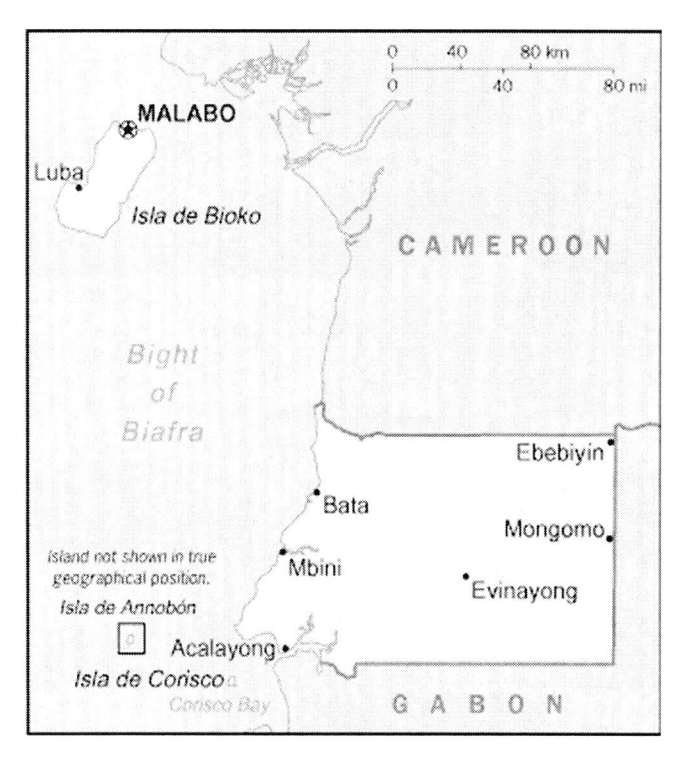

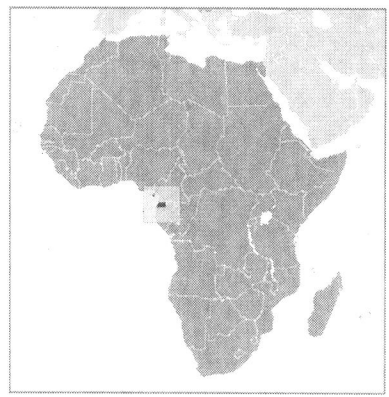

853. What is the official name of Equatorial Guinea?
    Republic of Equatorial Guinea
854. Which country borders Equatorial Guinea to the north?
    Cameroon
855. Which country borders Equatorial Guinea to the south and east?
    Gabon
856. Which water body borders Equatorial Guinea to the west?
    Gulf of Guinea
857. What is the motto of Equatorial Guinea?
    Unity, Peace, Justice
858. What is the national anthem of Equatorial Guinea?
    Let Us Tread the Path
859. What is the capital of Equatorial Guinea?
    Malabo (the largest city)
860. What are official languages of Equatorial Guinea?
    Spanish, French and Portuguese
861. What kind of government does Equatorial Guinea have?
    Presidential Republic
862. What was Spanish Guinea?
    An African colony of Spain that is the present day Equatorial Guinea
863. During which period did Spanish Guinea exist?
    August 11th, 1926– October 12th, 1968
864. What was the currency of Spanish Guinea?
    Spanish Guinea Peseta

865. What was the capital of Spanish Guinea?
Santa Isabel
866. Which town was renamed Santa Isabel by the Spanish?
Malabo
867. In which year was Santa Isabel renamed back to Malabo?
1973
868. What was the official name of Spanish Guinea?
Spanish Territories of the Gulf of Guinea
869. When did Equatorial Guinea gain independence from Span?
October 12th, 1968
870. What is the area of Equatorial Guinea?
10,828 sq mi / 28,051 km²
871. How long is the coastline of Equatorial Guinea?
184 mi / 296 km
872. What is the population of Equatorial Guinea?
633,441 (by 2009)
873. What is the currency of Equatorial Guinea?
Central African CFA Franc
874. What is the geographical feature of the terrain of Equatorial Guinea?
Coastal plains rise to interior hills; islands are volcanic
875. Equatorial Guinea comprises which two regions?
The Continental Region (Río Muni) and the Insular Region
876. Which small offshore islands are included in the Continental Region?
Corisco, Elobey Grande and Elobey Chico
877. Río Muni was ceded by Portugal to which country in 1778 in the Treaty of El Pardo?
Spain
878. Río Muni became a province of Spanish Guinea along with Bioko in which year?
1959
879. What is the largest island of the Gulf of Guinea, located in the Insular Region?
Bioko (formerly Fernando Po until the 70's)
880. How large is Bioko?
779 sq mi / 2,017 km²
881. What is the most remote island of the Gulf of Guinea, located in the Insular Region?
Annobón
882. How large is Annobón?
7 sq mi / 17 km²
883. Which island is the capital Malabo located?
Bioko
884. What are the five inhabitant islands in Equatorial Guinea?
Corisco, Elobey Grande, Elobey Chico, Annobón, and Bioko
885. What is the highest point of Equatorial Guinea?
Pico Basile (9,869 ft / 3,008 m)
886. What is the lowest point of Equatorial Guinea?
Atlantic Ocean (0 ft / 0 m)

887. How large is Monte Allen national Park, located in Centro Sur Province, which is a habitat to more than 105 species of mammals, 2300 species of birds, 65 species of reptiles, 55 species of amphibians and 65 species of aquatic lives?
618 sq mi / 1,600 km²
888. What are administrative divisions called in Equatorial Guinea?
Provinces
889. How many provinces does Equatorial Guinea have?
7
890. What is the climate of Equatorial Guinea?
Tropical; always hot, humid
891. What are the natural resources of Equatorial Guinea?
Petroleum, natural gas, timber, gold, bauxite, diamonds, tantalum, sand and gravel, clay
892. What are the natural hazards of Equatorial Guinea?
Violent windstorms; flash floods
893. What are the religions of Equatorial Guinea?
Nominally Christian and predominantly Roman Catholic, pagan practices
894. What are the ethnic groups of Equatorial Guinea?
Fang 85.7%, Bubi 6.5%, Mdowe 3.6%, Annobon 1.6%, Bujeba 1.1%, other 1.4%

Eritrea

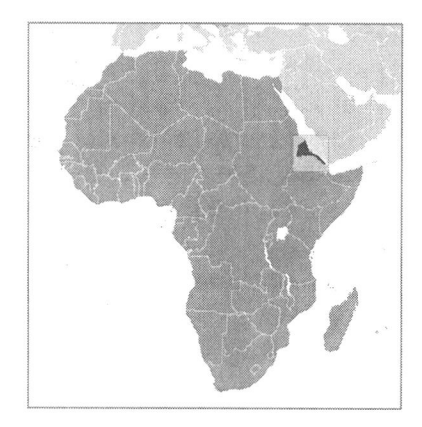

895. What is the official name of Eritrea?
State of Eritrea
896. Is Eritrea located in the Horn of Africa?

Yes
897. Which country borders Eritrea to the west?
Sudan
898. Which country borders Eritrea to the south?
Ethiopia
899. Which country borders Eritrea to the southeast?
Djibouti
900. The east and northeast of Eritrea have an extensive coastline on the Red Sea, directly across from which countries?
Saudi Arabia and Yemen
901. The Dahlak Archipelago is an island group located in which water body?
Red Sea
902. The Dahlak Archipelago consists of how many islands?
126 (2 large and 124 small)
903. Which island in the Dahlak Archipelago is the biggest and most populous?
Dahlak Kebir
904. Why have the pearl fisheries of the Dahlak Archipelago been famous since Roman times?
Because they produce a substantial number of pearls
905. The Hanish Islands are an island group in the Red Sea. Which country owns the larger islands while Eritrea owns the peripheral islands to the southwest of the larger islands?
Yemen
906. What is the national anthem of Eritrea?
Eritrea, Eritrea, Eritrea
907. What is the capital of Eritrea?
Asmara (the largest city)
908. What are official languages of Eritrea?
Arabic, English, and Tigrinya
909. What kind of government does Eritrea have?
Provisional government
910. On January 1st, 1890 Eritrea officially became a colony of which country?
Italy
911. Eritrea became part of which colony in 1936?
Italian East Africa
912. When did Eritrea gain independence from Italy?
November 1941
913. When did Eritrea gain independence from United Kingdom under UN Mandate?
1951
914. Which country made Eritrea its 14th province in 1962?
Ethiopia
915. When did Eritrea gain de facto independence from Ethiopia?
May 24th, 1991
916. When did Eritrea gain de jure independence from Ethiopia?
May 24th, 1993
917. What is the area of Eritrea?
45,405 sq mi / 117,600 km^2

918. How long is the coastline of Eritrea?
    1,388 mi / 2,234 km
919. What is the population of Eritrea?
    5,647,168 (by 2009)
920. What is the currency of Eritrea?
    Nakfa
921. What is the geographical feature of the terrain of Eritrea?
    Dominated by extension of Ethiopian north-south trending highlands, descending on the east to a coastal desert plain, on the northwest to hilly terrain and on the southwest to flat-to-rolling plains
922. Eritrea is virtually bisected by which mountain range?
    Great Rift Valley
923. Eritrea is a part of which depression?
    Afar Depression
924. What is the highest point of Eritrea?
    Emba Soira (9,902 ft / 3,018 m)
925. What is the lowest point of Eritrea?
    Near Kulul within the Afar Depression (-246 ft / -75 m)
926. Eritrea contains what national parks?
    Dahlak Marine National Park and Semenawi Bahri National Park
927. What are administrative divisions called in Eritrea?
    Regions
928. How many regions does Eritrea have?
    6
929. What is the climate of Eritrea?
    Hot, dry desert strip along Red Sea coast; cooler and wetter in the central highlands (up to 2 ft / 61 cm of rainfall annually, heaviest June to September); semiarid in western hills and lowlands
930. What are the natural resources of Eritrea?
    Gold, potash, zinc, copper, salt, possibly oil and natural gas, fish
931. What are the natural hazards of Eritrea?
    Frequent droughts; locust swarms
932. What are the religions of Eritrea?
    Muslim, Coptic Christian, Roman Catholic, Protestant
933. What are the ethnic groups of Eritrea?
    Tigrinya 50%, Tigre and Kunama 40%, Afar 4%, Saho (Red Sea coast dwellers) 3%, other 3%

Ethiopia

934. What is the official name of Ethiopia?
    Federal Democratic Republic of Ethiopia
935. Which country borders Ethiopia to the west?
    Sudan
936. Which countries border Ethiopia to the north?
    Djibouti and Eritrea

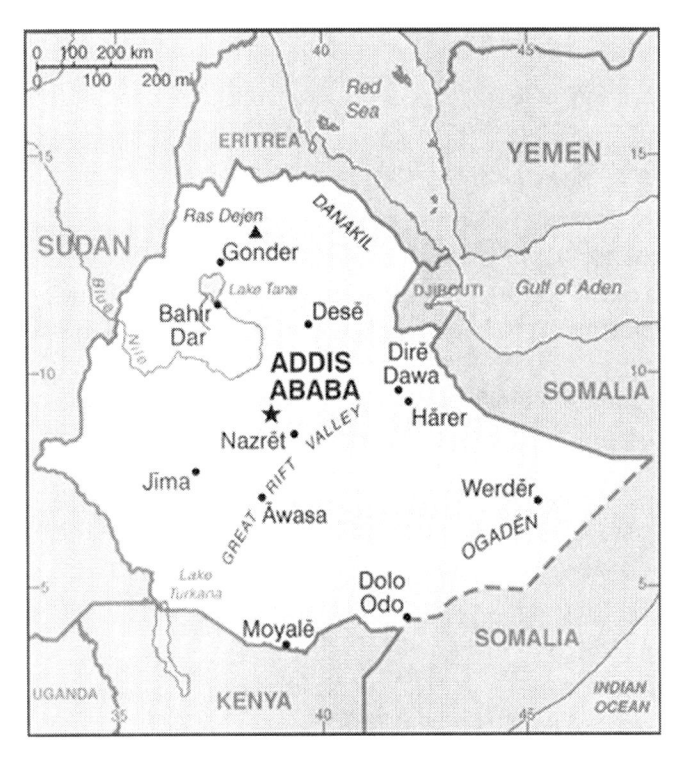

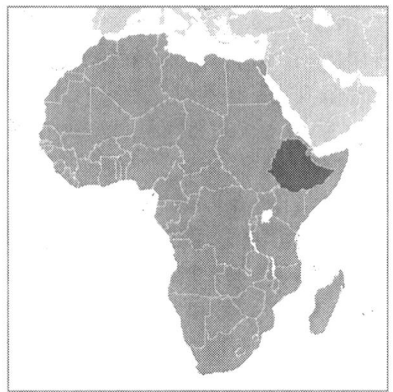

937. Which country borders Ethiopia to the east?
    Somalia
938. Which country borders Ethiopia to the south?
    Kenya
939. What is the national anthem of Ethiopia?
    March Forward, Dear Mother Ethiopia
940. What is the capital of Ethiopia?
    Addis Ababa (the largest city)
941. What is the official language of Ethiopia?
    Amharic
942. What kind of government does Ethiopia have?
    Federal Parliamentary republic
943. When was Ethiopia established?
    10th century BCE
944. D'mt was a kingdom located in Eritrea and northern Ethiopia that existed during which period?
    8th century BCE–7th century BCE
945. When did Proto-Aksum period exist?
    5th century BCE–1st century BCE
946. When did Aksum (also called Aksumite Empire, Axumite Empire, the Kingdom of Aksum, or the Kingdom of Axum) exist?
    1st century BCE–10th century CE
947. When did Aksum, located in Eritrea and northern Ethiopia, use the name Ethiopia?
    4th century CE

948. Aksum was the first major empire to convert to which religion?
Christianity
949. Aksum was the alleged resting place of what very famous religious article?
The Ark of the Covenant, a vessel described in the Bible as containing the Tablets of Stone on which the Ten Commandments were inscribed, along with Aaron's rod and manna?
950. The Queen of Sheba, the ruler of the kingdom mentioned in the Old Testament and the Koran, allegedly lived where in Ethiopia?
Aksum
951. Who was Gudit in 10th century CE?
A semi-legendary non-Christian queen who destroyed the remnants of Aksum
952. Which dynasty ruled Ethiopia from approximately 1137 to 1270?
Zagwe Dynasty
953. When was the First Solomonic period?
1270–1527
954. Invasion of Gragn during 1527–1543 brought three-quarters of Ethiopia under the power of which religion?
Islam
955. Habesh (an Ottoman province) during 16th–17th century included which present-day countries?
Northern parts of Ethiopia, Eritrea, Djibouti, and Somalia
956. When did Gondarine dynasty exist?
1606–1755
957. What kind of period is the Zemene Mesafint during 1755–1855?
A period with a lot of conflicts between warlords
958. What is the kind of period of Ethiopia during 1855–1936?
Modernization
959. The Second Italo–Abyssinian War (also called Second Italo-Ethiopian War) was a brief colonial war that started in October 1935 and ended in May 1936. The war was fought between the Ethiopian Empire (also called Abyssinia) and which country?
Italy
960. Italian East Africa was consisted of which countries?
Ethiopia, Italian Somaliland and Eritrea
961. When did Italian East Africa exist?
1936–1941
962. What was the capital of Italian East Africa?
Addis Ababa
963. What was the East African Campaign?
A series of battles fought in East Africa during World War II (June 10th, 1940–November 27th, 1941) between the British Empire, the British Commonwealth of Nations, and several allies, against the forces of the Italian Empire
964. East African Campaign was fought in which countries?
Sudan, British Somaliland, Kenya, Eritrea, Italian Somaliland, Ethiopia
965. What was the result of the East African Campaign?
Decisive allied victory, the fall of Italian East Africa

966. What was the Italian guerrilla war in Ethiopia fought from the summer of 1941 to the autumn of 1943?
A war fought by remnants of Italian troops in Italian East Africa after the East African Campaign
967. Ethiopia went through the Second Modernization from 1941 to when?
1974
968. Ethiopia was involved in the independence war of which country during 1961–1991?
Eritrea
969. How did the Ethiopian Civil War begin?
The Marxist Derg staged a coup d'état against Emperor Haile Selassie on September 12th, 1974
970. How did the Ethiopian Civil War end?
The Ethiopian People's Revolutionary Democratic Front (EPRDF), a coalition of rebel groups, overthrew the government in 1991
971. What was the Ethiopian crackdown in Ogaden during 2007–2008?
A campaign involving the Ethiopian Army on the offensive against the rebel Ogaden National Liberation Front (ONLF)
972. What was the result of the Ethiopian crackdown in Ogaden?
Indecisive; Ethiopians claimed victory, ONLF denied that
973. How did the War in Somalia begin?
The war officially began shortly before July 20, 2006 when U.S. backed Ethiopian troops invaded Somalia to prop up Somali Transitional Federal Government (TFG) in Baidoa
974. How did the War in Somalia end?
As of January 2009, Ethiopian troops withdrew from Somalia following two year insurgencies which lead to loss of territory and effectiveness of the TFG and a power sharing deal between Islamists splinter group. The war ended on January 30th, 2009, one day before TFG had the elected President
975. What was the result of the War in Somalia?
Islamist victory; political victory for the Alliance for the Re-liberation of Somalia
976. Besides Ethiopia, how many countries in Africa have its own alphabet?
0
977. What is the area of Ethiopia?
426,371 sq mi / 1,104,300 km^2
978. How long is the coastline of Ethiopia?
0 mi / 0 km
979. What is the population of Ethiopia?
85,237,338 (by 2009)
980. What is Ethiopia's rank by population in Africa?
2nd
981. What is the currency of Ethiopia?
Birr
982. What is the geographical feature of the terrain of Ethiopia?
High plateau with central mountain range divided by Great Rift Valley
983. What is the highest point of Ethiopia?
Ras Dejen (also called Ras Dashen, 14,872 ft / 4,533 m)

984. Ras Dejen is located in which mountain range?
Semien Mountains
985. Ras Dejen is the eastern peak of the rim of an enormous volcano. What is the western peak of the rim?
Mount Biuat (14,797 ft / 4,510 m)
986. Ras Dejen and Mount Biuat are separated by the valley of which river?
The Meshaha River
987. What is the Ethiopian Highlands?
A rugged mass of mountains in Ethiopia, Eritrea, and Somalia
988. Why is the Ethiopian Highlands called the Roof of Africa?
Because of its height and large area
989. The Ethiopian Highlands is divided into northwestern and southeastern portions by what?
Great Rift Valley
990. Semien Mountains is in the northwestern portion or southeastern portion of the Ethiopian Highlands?
Northwestern portion
991. What are notable animals in Semien Mountains?
Walia Ibex, Gelada Baboon, Caracal and the Ethiopian Wolf
992. Why is Simien Mountains National Park in danger?
Because of serious population declines of some of its characteristic native species
993. Which lake, as the source of the Blue Nile, lies in the northwestern portion of Ethiopian Highlands?
Lake Tana
994. What is the largest lake in Ethiopia?
Lake Tana (832 sq mi / 2,156 km²)
995. What are primary inflows of Lake Tana?
Lesser Abay, Kilti River, Magech River, Reb River, and Gumara River
996. What is the primary outflow of Lake Tana?
Blue Nile
997. What is the surface elevation of Lake Tana?
5,866 ft / 1,788 m
998. What are the important islands in Lake Tana?
Tana Qirqos, Daga Island, Dek Island, and Meshralia
999. Why are islands in Lake Tana important?
Because remains of ancient Ethiopian emperors and treasures of the Ethiopian Church are kept in the isolated island monasteries
1000. Which mountain range is in the southeastern portion of the Ethiopian Highlands?
The Bale Mountains
1001. The largest group of which animal is in the Bale Mountains?
Ethiopian Wolves
1002. Which peak in the Bale Mountains is the second-highest mountain in Ethiopia?
Tullu Demtu (14,360 ft / 4,377 m)
1003. Bale Mountains contains which three distinct ecoregions?

The northern plains, bush and woods; the central Sanetti Plateau with an average elevation of over 13,123 ft / 4000 m; and the southern Harenna Forest, known for its mammals, amphibians and birds including many endemic species

1004. What is the Bale Mountains National Park?
A national park in the Oromia Region of southeast Ethiopia

1005. How large is the Bale Mountains National Park?
849 sq mi / 2,200 km²

1006. As part of the Bale National Park, how high is the Mount Batu?
14,131 ft / 4,307 m

1007. Which peak of the Mount Batu is higher, Tinnish Batu (Little Batu), or Tilliq Batu (Big Batu)?
Tinnish Batu

1008. What is the Awash River (also called Hawash River)?
A major river of Ethiopia that entirely contained within the boundaries of Ethiopia

1009. How long is the Awash River?
746 mi / 1,200 km

1010. What are tributaries of the Awash River?
Logiya River, Mille River, Borkana River, Ataye River, Hawadi River, Kabenna River and Durkham River

1011. The Awash River begins with Lake Gargori and end with which lake?
Lake Abbe

1012. Which highway passes through the Awash National Park?
The Addis Ababa - Dire Dawa

1013. What animals are in the Awash National Park?
East African Oryx, Soemmerring's Gazelle, Dik-dik, the lesser and greater Kudus, and over 350 species of native birds

1014. Lake Abbe is located in which depression?
Afar Depression

1015. Lake Abbe is one of a chain of six connected salt lakes. What are other five lakes?
Gargori, Laitali, Gummare, Bario and Afambo

1016. Lake Abbe is known for its limestone chimneys, from which steam vents. How high are these chimneys?
164 ft / 50 m

1017. What is the lowest point of Ethiopia?
Afar Depression (-410 ft / -125 m)

1018. What is Ethiopian xeric grasslands and shrublands ecoregion?
A semi-desert strip on or near the Red Sea and the Gulf of Oman coasts in Eritrea, Ethiopia, Djibouti and Somalia

1019. Ethiopia has many spa towns, which are a result of Ethiopia's abundance of what natural occurrence?
Hot springs

1020. What are administrative divisions called in Ethiopia?
Ethnically based states

1021. How many states does Ethiopia have?
9 (plus 2 self-governing administrations)

1022. What is the climate of Ethiopia?

Tropical monsoon with wide topographic-induced variation
1023. What are the natural resources of Ethiopia?
Small reserves of gold, platinum, copper, potash, natural gas, hydropower
1024. Which three major crops are believed to have originated in Ethiopia?
Coffee, grain sorghum, and castor bean
1025. What are the natural hazards of Ethiopia?
Geologically active Great Rift Valley susceptible to earthquakes, volcanic eruptions; frequent droughts
1026. What are the religions of Ethiopia?
Christian 60.8% (Orthodox 50.6%, Protestant 10.2%), Muslim 32.8%, traditional 4.6%, other 1.8%
1027. What are the ethnic groups of Ethiopia?
Oromo 32.1%, Amara 30.1%, Tigraway 6.2%, Somalie 5.9%, Guragie 4.3%, Sidama 3.5%, Welaita 2.4%, other 15.4%

Gabon

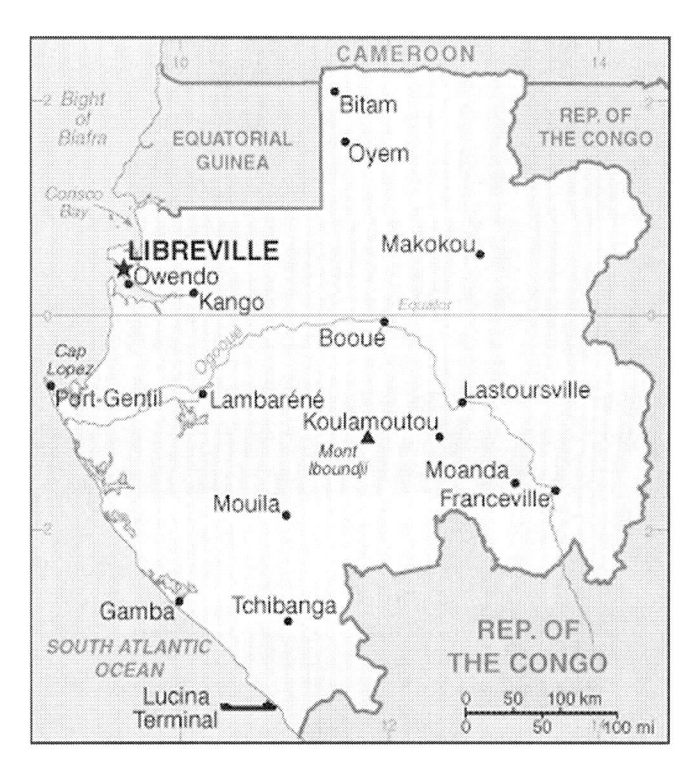

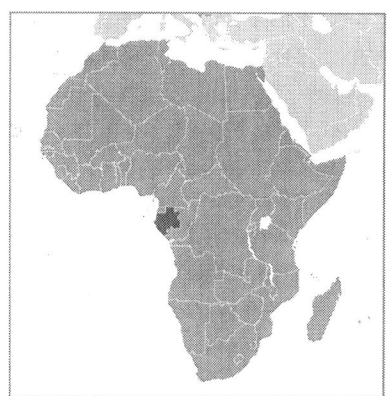

1028. What is the official name of Gabon?
Gabonese Republic
1029. Which country borders Gabon to the northwest?
Equatorial Guinea
1030. Which country borders Gabon to the northwest?
Cameroon
1031. Which country borders Gabon to the east and south?

Republic of the Congo
1032. Which water body borders Gabon to the west?
Gulf of Guinea
1033. Is Gabon located on the equator?
Yes
1034. What is the motto of Gabon?
Union, Work, Justice
1035. What is the national anthem of Gabon?
The Concorde
1036. What is the capital of Gabon?
Libreville (the largest city)
1037. What is the official language of Gabon?
French
1038. What kind of government does Gabon have?
Republic
1039. In which year did Gabon become one of the four territories of French Equatorial Africa?
1910
1040. When did Gabon gain independence from France?
August 17th, 1960
1041. What is the area of Gabon?
103,347 sq mi / 267,745 km²
1042. How long is the coastline of Gabon?
550 mi / 885 km
1043. What is the population of Gabon?
1,514,993 (by 2009)
1044. What is the currency of Gabon?
Central African CFA Franc
1045. What is the geographical feature of the terrain of Gabon?
Narrow coastal plain; hilly interior; savanna in east and south
1046. What is the highest point of Gabon?
Mont Iboundji (5,167 ft / 1,575 m)
1047. What is the lowest point of Gabon?
Atlantic Ocean (0 ft / 0 m)
1048. What is the longest river in Gabon?
The Ogooué (also called Ogowe, 750 mi / 1,200 km)
1049. The Ogooué drains nearly the entire country of Gabon with some tributaries reaching into which other countries?
The Republic of the Congo, Cameroon, and Equatorial Guinea
1050. Gabon has three karst areas where there are hundreds of caves located in the dolomite and limestone rocks. What are some names of the caves?
Grotte du Lastoursville, Grotte du Lebamba, Grotte du Bongolo, and Grotte du Kessipougou
1051. How large is the Loango National Park, located in western Gabon, which is home to dolphins, whales and many land mammals?
598 sq mi / 1,550 km²
1052. The Loango National Park includes part of Iguéla Lagoon. How large is Iguéla Lagoon?

85 sq mi / 220 km²
1053. The Ndogo Lagoon, located in the southwestern part of Gabon, has a very large density of islands. How many islands are there in the lagoon?
About 350
1054. How large is Akanda National Park, located in northeastern Gabon, which consists of marine mangroves, forest and savannah?
208 sq mi / 540 km²
1055. How large is Batéké Plateau National Park, located in southeastern Gabon, which is famous for the diversity of birds with an idyllic landscape?
792 sq mi / 2,050 km²
1056. How large is Mont Birougou National Park, located in the southern Gabon, which is the source of the Myanga River, Lolo River and Onoy River?
266 sq mi / 690 km²
1057. How large is Crystal Mountains National Park, located in northwestern Gabon, which is home to many plant species?
463 sq mi / 1,200 km²
1058. How large is Ivindo National Park, located in east-central Gabon, which contains the Mingouli, Koungou and Djidji Falls?
1,158 sq mi / 3,000 km²
1059. How large is Lopé National Park, located in central Gabon, which is said to be the most popular safari destination in Gabon?
1,896 sq mi / 4,910 km²
1060. How large is Mayumba National Park, a marine park in southwestern Gabon, which shelters 37 miles (60 km) of the most important leatherback turtle nesting beach on Earth?
336 sq mi / 870 km²
1061. How large is Minkébé National Park, located in the extreme northeastern of Gabon, which is home to some very rare mammals such as the bongo and the giant forest pig?
2,923 sq mi / 7,570 km²
1062. How large is Moukalaba-Doudou National Park, located in south western Gabon, which includes Nyanga plains, Doudou Mountains and large papyrus near the Nyanga River?
1,737 sq mi / 4,500 km²
1063. How large is Mwangné National Park, located along the Congo border, which is home to large populations of elephants and rare animals like bongos and sitatungas?
448 sq mi / 1,160 km²
1064. How large is Pongara National Park, located south of Libreville, which contains large areas of mangrove, forest and savannah?
359 sq mi / 929 km²
1065. How large is Waka National Park, located in central Gabon, which includes the over 62 miles (100 km) long Ikobe-Ikoy-Onoy rift valley?
409 sq mi / 1,060 km²
1066. What are administrative divisions called in Gabon?
Provinces
1067. How many provinces does Gabon have?
9
1068. What is the climate of Gabon?

Tropical; always hot, humid

1069. What are the natural resources of Gabon?
Petroleum, natural gas, diamond, niobium, manganese, uranium, gold, timber, iron ore, hydropower

1070. What are the religions of Gabon?
Christian 55%-75%, animist, Muslim less than 1%

1071. What are the ethnic groups of Gabon?
Bantu tribes, including four major tribal groupings (Fang, Bapounou, Nzebi, Obamba); other Africans and Europeans, 154,000, including 10,700 French and 11,000 persons of dual nationality

Gambia, The

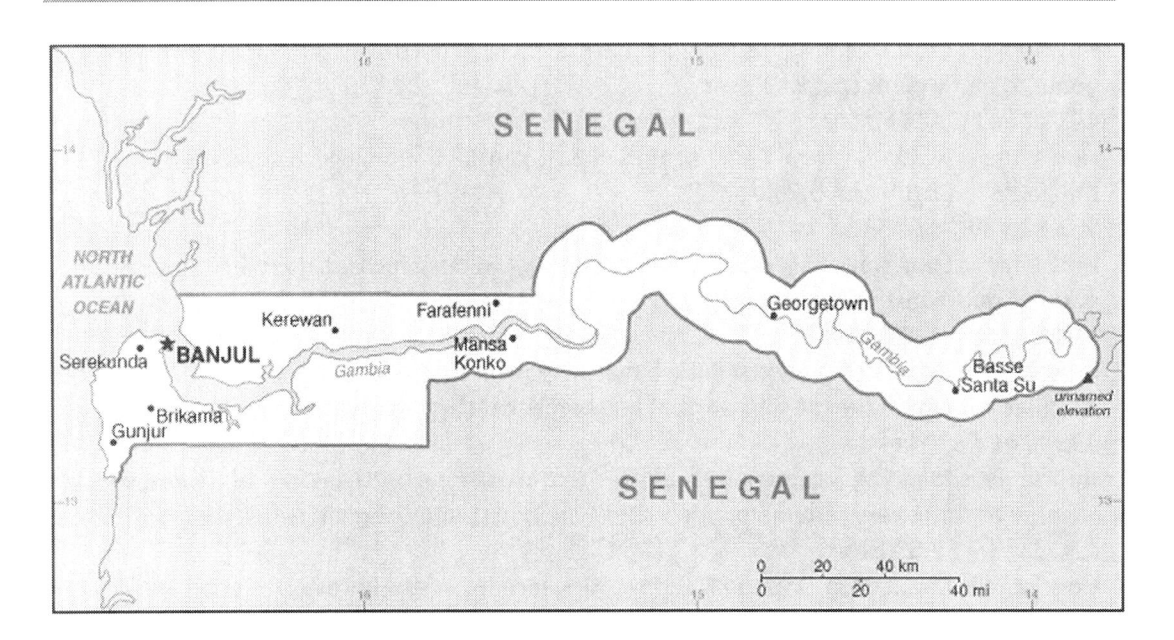

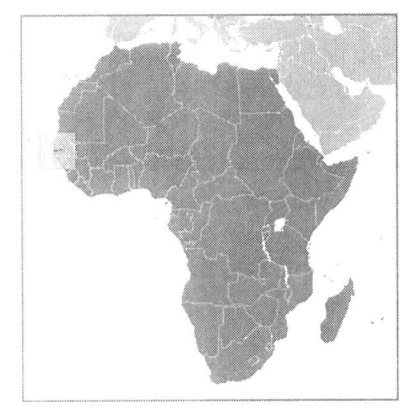

1072. What is the official name of Gambia?
Republic of The Gambia

1073. Which country borders Gambia to the north, east, and south?
Senegal
1074. Which water body borders Gambia to the west?
Atlantic Ocean
1075. Gambia consists of little more than the downstream half of which river and its two banks?
The Gambia River
1076. How long is the Gambia River?
700 mi / 1,130 km
1077. What is Gambia's rank by size on the continent Africa?
Smallest
1078. What is the motto of Gambia?
Progress, Peace, Prosperity
1079. What is the national anthem of Gambia?
For The Gambia Our Homeland
1080. What is the capital of Gambia?
Banjul
1081. What is the largest city of Gambia?
Serrekunda
1082. What is the official language of Gambia?
English
1083. What kind of government does Gambia have?
Republic
1084. During the late 17th and throughout the 18th century, which two countries constantly struggled for political and commercial supremacy in the regions of the Senegal and Gambia Rivers?
United Kingdom and France
1085. Which treaty gave United Kingdom possession of The Gambia?
The 1783 Treaty of Paris
1086. During World War II, Gambian was on which side?
The Allies
1087. When did Gambia gain independence from United Kingdom?
February 18th, 1965
1088. What is the area of Gambia?
4,007 sq mi / 10,380 km^2
1089. How long is the coastline of Gambia?
50 mi / 80 km
1090. What is the population of Gambia?
1,782,893 (by 2009)
1091. What is the currency of Gambia?
Dalasi
1092. What is the geographical feature of the terrain of Gambia?
Flood plain of the Gambia River flanked by some low hills
1093. What is the highest point of Gambia?
Unnamed location (174 ft / 53 m)
1094. What is the lowest point of Gambia?

        Atlantic Ocean (0 ft / 0 m)
1095. Which national park is Gambia's first designated wildlife reserve?
Abuko National Park (0.4 sq mi / 1 km^2)
1096. How large is Bao Bolong Wetland Reserve, located in North Bank Division of Gambia?
85 sq mi / 220 km²
1097. How large is Kiang West National Park, located in Lower River Division of Gambia, which is mostly composed of Guinea savanna and dry deciduous woodland?
44 sq mi / 115 km^2
1098. How large is Niumi National Park, located on the coastal strip that is north of the Gambia River, which has one of the last untouched mangrove stands on the West African Coast north of the equator?
19 sq mi / 49 km^2
1099. How large is River Gambia National Park, located in Central River Division of Gambia, which includes Baboon Islands?
2.3 sq mi / 5.9 km^2
1100. What are administrative divisions called in Gambia?
Divisions
1101. How many divisions does Gambia have?
5 (plus 1 city)
1102. What is the climate of Gambia?
Tropical; hot, rainy season (June to November); cooler, dry season (November to May)
1103. What are the natural resources of Gambia?
Fish, titanium (rutile and ilmenite), tin, zircon, silica sand, clay, petroleum
1104. What are the natural hazards of Gambia?
Drought (rainfall has dropped by 30% in the last 30 years)
1105. What are the religions of Gambia?
Muslim 90%, Christian 8%, indigenous beliefs 2%
1106. What are the ethnic groups of Gambia?
Mandinka 42%, Fula 18%, Wolof 16%, Jola 10%, Serahuli 9%, non-African 1%, other 4%

Ghana

1107. What is the official name of Ghana?
Republic of Ghana
1108. Which country borders Ghana to the west?
Côte d'Ivoire
1109. Which country borders Ghana to the north?
Burkina Faso
1110. Which country borders Ghana to the east?
Togo
1111. Which water body borders Ghana to the south?
Gulf of Guinea
1112. What is the motto of Ghana?
Freedom and Justice

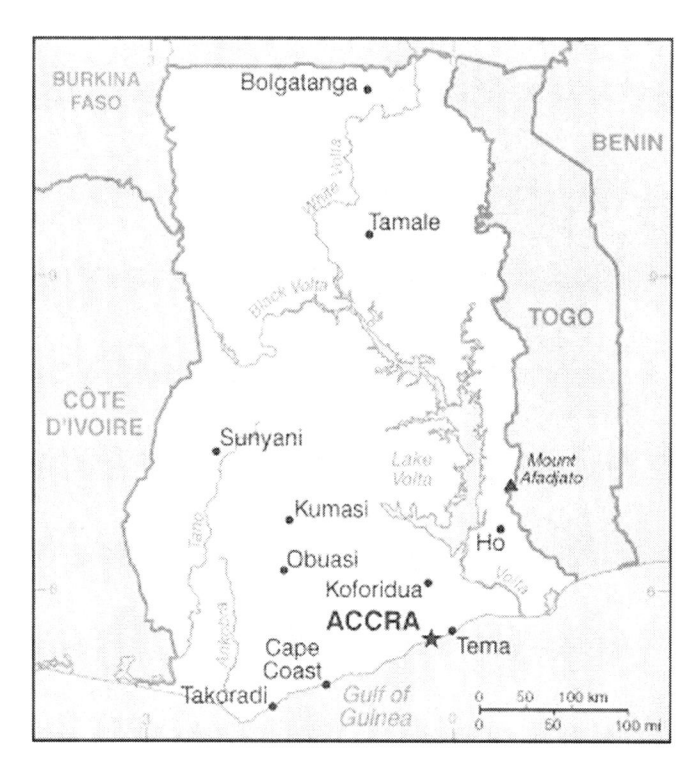

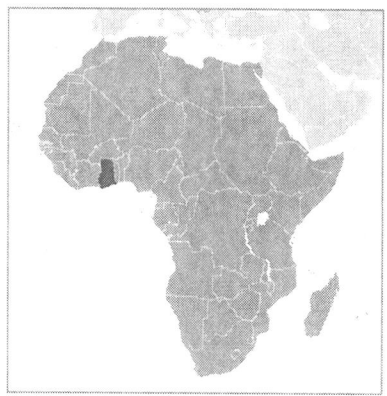

1113. What is the national anthem of Ghana?
God Bless Our Homeland Ghana
1114. What is the capital of Ghana?
Accra (the largest city)
1115. What is the official language of Ghana?
English
1116. What kind of government does Ghana have?
Constitutional democracy
1117. Was Ghana part of the Ghana Empire (also called Wagadou Empire) during 790–1076?
No
1118. What was the Ashanti Empire (also called Asante Empire, Ashanti Confederacy or Asanteman) during 1670–1902?
A pre-colonial West African state of what is now the Ashanti Region in Ghana
1119. The Ashanti Empire stretched from central Ghana to which present-day countries?
Togo and Côte d'Ivoire
1120. What was the capital of the Ashanti Empire?
Kumasi
1121. What was the language of the Ashanti Empire?
Twi
1122. What was the name of Ghana as a British colony during 1821-1957?
Gold Coast
1123. What was the capital of Gold Coast during 1821-1877?
Cape Coast
1124. What was the capital of Gold Coast during 1877-1957?

Accra
1125. When did British Togoland Integrate with Gold Coast?
December 13th, 1956
1126. What was British Togoland?
A League of Nations Class B mandate in Africa, formed by the splitting of German protectorate Togoland into French Togoland and British Togoland
1127. What was the capital of British Togoland?
Ho
1128. When did Ghana gain independence from United Kingdom?
March 6th, 1957
1129. What was the significance of Ghana's independence?
It was the first sub-Saharan African country to gain its independence
1130. What is the area of Ghana?
92,098 sq mi / 238,535 km^2
1131. How long is the coastline of Ghana?
335 mi / 539 km
1132. What is the population of Ghana?
23,832,495 (by 2009)
1133. What is the currency of Ghana?
Ghanaian Cedi
1134. What is the geographical feature of the terrain of Ghana?
Mostly low plains with dissected plateau in south-central area
1135. What is the highest point of Ghana?
Mount Afadjato (2,887 ft / 880 m)
1136. What is the lowest point of Ghana?
Atlantic Ocean (0 ft / 0 m)
1137. The Greenwich Meridian passes through which city of Ghana?
Tema
1138. Ghana is geographically closer to the "centre" of the world than any other country. Where is the actual centre, (0°, 0°)?
In the Atlantic Ocean approximately 382 mi (614 km) south of Accra, Ghana, in the Gulf of Guinea
1139. What is the Lake Volta's rank by surface area of man-made lakes in the world?
1st (3,283 sq mi / 8,502 km^2)
1140. What is Lake Volta's rank by water volume of man-made lakes in the world?
4th (57,000 sq mi / 148 km^3)
1141. What are the inflows of Lake Volta?
White Volta River and Black Volta River
1142. What is the outflow of the Lake Volta?
Volta River
1143. What is the longest river in Ghana?
Volta River
1144. What is Akosombo Dam?
A hydroelectric dam in southeastern Ghana in the Akosombo gorge on the Volta River
1145. In which year did the construction of the Akosombo Dam begin?

1961
1146. In which year did Akosombo Dam open?
1965
1147. What is the result of Akosombo Dam?
Lake Volta
1148. Which national park lies on part of Lake Volta's west shore?
The Digya National Park
1149. Which national park is Ghana's largest wildlife refuge?
The Mole National Park (1,869 sq mi / 4,840 km^2)
1150. How large is Bia National Park, located in Western Region of Ghana, which is named after the Bia River which drains the area?
217 sq mi / 563 km²
1151. How large is Bui National Park, located in Brong-Ahafo Region and Northern Region of Ghana, which is composed of open woodland savannah and grassland?
695 sq mi / 1,800 km^2
1152. How large is Kakum National Park, located in Central Region of Ghana, which is home to tropical plants and rare animals, including the very rare and endangered Mona-Meerkat, as well as pygmy elephants, forest buffalo, and civet cats?
135 sq mi / 350 km^2
1153. Kakum National Park has a long hanging bridge over 1,080 ft (330 m) long in the forest canopy level. What is the bridge called?
The Canopy Walkway
1154. What is Cape Three Points?
A small peninsula that forms the southernmost tip of Ghana. It is known as the "land nearest nowhere" because it is the land nearest a location in the sea which is (0°, 0°)
1155. How long is the Ankobra River in southern Ghana?
120 mi / 190 km
1156. The Ankobra River is fed by which rivers?
Fure River, Mansi River, Nini River, and Bonsa River
1157. How long is the Tano River (also called Tanoé River) in ghana?
249 mi / 400 km
1158. The Atakora Mountains in Benin is called what in Ghana?
Akwapim Hills
1159. What are administrative divisions called in Ghana?
Regions
1160. How many regions does Ghana have?
10
1161. What is the climate of Ghana?
Tropical; warm and comparatively dry along southeast coast; hot and humid in southwest; hot and dry in north
1162. What are the natural resources of Ghana?
Gold, timber, industrial diamonds, bauxite, manganese, fish, rubber, hydropower, petroleum, silver, salt, limestone
1163. What are the natural hazards of Ghana?
Dry, dusty, northeastern harmattan winds occur from January to March; droughts

1164. What are the religions of Ghana?
    Christian 68.8% (Pentecostal/Charismatic 24.1%, Protestant 18.6%, Catholic 15.1%, other 11%), Muslim 15.9%, traditional 8.5%, other 0.7%, none 6.1%
1165. What are the ethnic groups of Ghana?
    Akan 45.3%, Mole-Dagbon 15.2%, Ewe 11.7%, Ga-Dangme 7.3%, Guan 4%, Gurma 3.6%, Grusi 2.6%, Mande-Busanga 1%, other tribes 1.4%, other 7.8%

Guinea

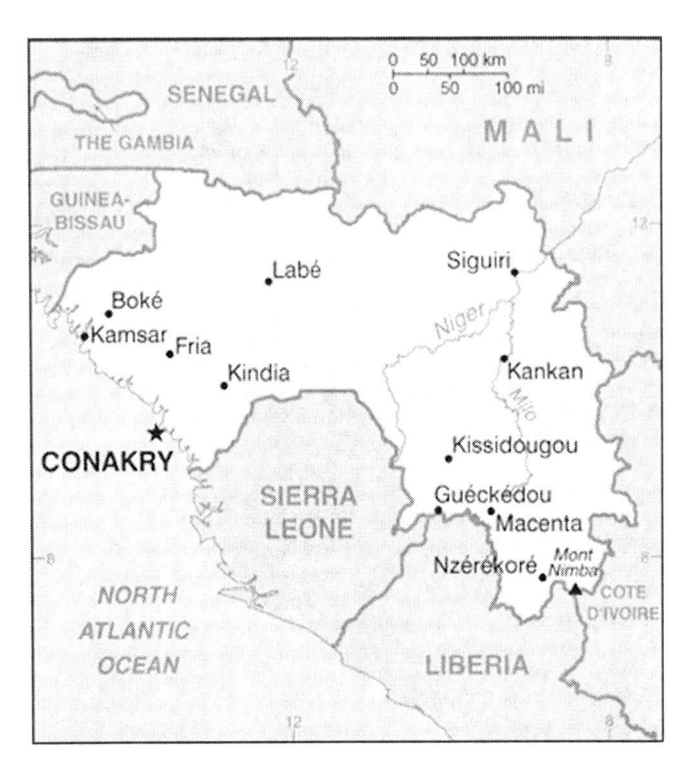

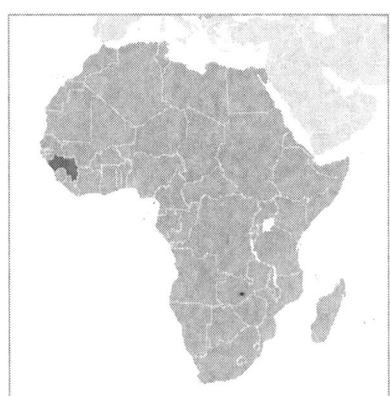

1166. What is the official name of Guinea?
    Republic of Guinea
1167. Which water body borders Guinea to the west?
    Atlantic Ocean
1168. Which country borders Guinea to the west?
    Guinea-Bissau
1169. Which country borders Guinea to the north and northeast?
    Mali
1170. Which country borders Guinea to the north?
    Senegal
1171. Which country borders Guinea to the southeast?
    Côte d'Ivoire
1172. Which country borders Guinea to the southwest?
    Sierra Leone
1173. Which country borders Guinea to the south?

Liberia
1174. Which river runs through Guinea, providing both water and irregular transportation?
The Niger River
1175. What is the motto of Guinea?
Work, Justice, Solidarity
1176. What is the national anthem of Guinea?
Freedom
1177. What is the capital of Guinea?
Conakry (the largest city)
1178. Guinea is sometimes called what to distinguish it from its neighbor Guinea-Bissau?
Guinea-Conakry
1179. What is the official language of Guinea?
French
1180. What kind of government does Guinea have?
Military junta
1181. Was northern Guinea part of Mali Empire during 1230s–1600s?
Yes
1182. What was the Kingdom of Fouta Djallon (also called Kingdom of Fuuta Jallon and the Timbo Almamate) during 1725–1896?
A pre-colonial West African state based in the Fouta Djallon highlands of the present-day Guinea
1183. What was the Wassoulou Empire (also called Mandinka Empire) during 1878–1898?
A short-lived empire in the predominately Malinké area of what is the present-day upper Guinea and southwestern Mali (Wassoulou)
1184. When was French Guinea established?
1894
1185. What was the currency of French Guinea during 1903–1945?
French West African Franc
1186. What was the currency of French Guinea during 1945–1958?
CFA franc
1187. When did Guinea gain independence from France?
October 2nd, 1958
1188. What is the area of Guinea?
94,926 sq mi / 245,857 km^2
1189. How long is the coastline of Guinea?
199 mi / 320 km
1190. What is the population of Guinea?
10,057,975 (by 2009)
1191. What is the currency of Guinea?
Guinean Franc
1192. What is the geographical feature of the terrain of Guinea?
Generally flat coastal plain, hilly to mountainous interior
1193. Guinea is divided into how many geographic regions?
4

1194. What is the geographic region name of the lowland belt running north to south behind the coast?
Lower Guinea
1195. Lower Guinea is part of which ecoregion?
Guinean forest-savanna mosaic
1196. What is the Guinean forest-savanna mosaic?
An ecoregion of West Africa, a band of interlaced forest, savanna, and grassland running east to west and a transitional habitat between the rain forests of the Guinean-Congolian region and the dry savannas of Sudan
1197. What is the geographic region name of the pastoral Fouta Djallon highlands?
Middle Guinea
1198. What is Fouta Djallon?
A highland region in the center of Guinea
1199. What is the geographic region name of the northern savanna?
Upper Guinea
1200. What is the geographic region name of the southeastern rain-forest region?
Forest Guinea
1201. There are 22 West African rivers that have their origins in Guinea. Please name three…
Niger River, Gambia River and Senegal River
1202. What is the highest point of Guinea?
Mount Nimba (5,748 ft / 1,752 m)
1203. Mount Nimba Strict Nature Reserve is a national park and UNESCO World Heritage Site in Guinea and which other country?
Côte d'Ivoire
1204. What is the lowest point of Guinea?
Atlantic Ocean (0 ft / 0 m)
1205. Haut Niger National Park (also called National Park of the Upper Niger, 2,317 sq mi / 6,000 km^2), located in the northeastern Guinea, comprises which two zones?
The core protected zone (232 sq mi / 600 km^2) and a buffer zone in which local people are encouraged to use the resources in a sustainable way
1206. National Park of the Niokolo-Badiar, home to antelopes, monkeys, lions, and leopards during the dry season, is a savanna along the border with which country?
Senegal
1207. What are administrative divisions called in Guinea?
Prefectures
1208. How many prefectures does Guinea have?
33 (plus 1 special zone)
1209. What is the climate of Guinea?
Generally hot and humid; monsoonal-type rainy season (June to November) with southwesterly winds; dry season (December to May) with northeasterly harmattan winds
1210. What are the natural resources of Guinea?
Bauxite, iron ore, diamonds, gold, uranium, hydropower, fish, salt
1211. What are the natural hazards of Guinea?
Hot, dry, dusty harmattan haze may reduce visibility during dry season
1212. What are the religions of Guinea?

Muslim 85%, Christian 8%, indigenous beliefs 7%
1213. What are the ethnic groups of Guinea?
Peuhl 40%, Malinke 30%, Soussou 20%, smaller ethnic groups 10%

Guinea-Bissau

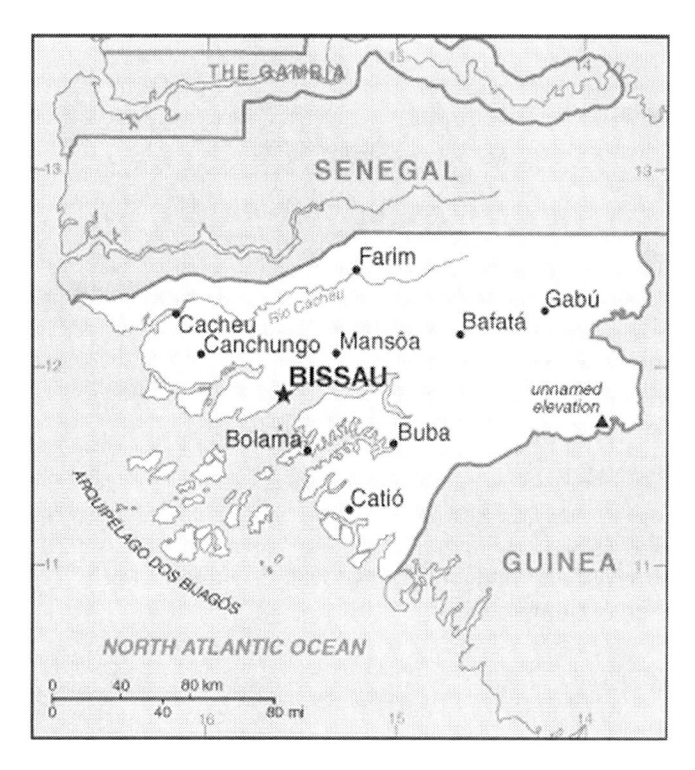

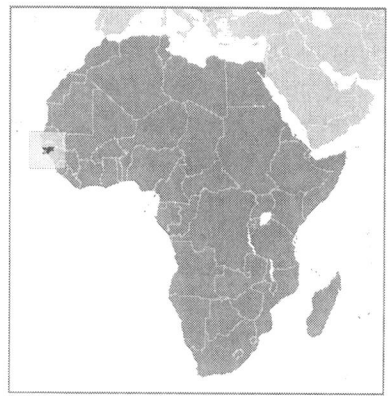

1214. What is the official name of Guinea-Bissau?
Republic of Guinea-Bissau
1215. Which country borders Guinea-Bissau to the north?
Senegal
1216. Which country borders Guinea-Bissau to the southeast?
Guinea
1217. Which water body borders Guinea-Bissau to the west?
Atlantic Ocean
1218. What is the motto of Guinea-Bissau?
Unity, Struggle, Progress
1219. What is the national anthem of Guinea-Bissau?
This is Our Well-Beloved Motherland
1220. What is the capital of Guinea-Bissau?
Bissau (the largest city)
1221. What is the official language of Guinea-Bissau?
Portuguese
1222. What kind of government does Guinea-Bissau have?
Semi-presidential republic

1223. What was the name for present-day Guinea-Bissau from 1446 to September 10, 1974?
Portuguese Guinea
1224. When did Guinea-Bissau declare independence from Portugal?
September 24th, 1973
1225. When was the independence of Guinea-Bissau recognized?
September 10th, 1974
1226. What is the area of Guinea-Bissau?
13,948 sq mi / 36,125 km^2
1227. How long is the coastline of Guinea-Bissau?
217 mi / 350 km
1228. What is the population of Guinea-Bissau?
1,533,964 (by 2009)
1229. What is the currency of Guinea-Bissau?
West African CFA franc
1230. What is the geographical feature of the terrain of Guinea-Bissau?
Mostly low coastal plain rising to savanna in east
1231. What is the highest point of Guinea-Bissau?
Unnamed location in the northeast corner of the country (984 ft / 300 m)
1232. What is the lowest point of Guinea-Bissau?
Atlantic Ocean (0 ft / 0 m)
1233. Guinea-Bissau contains what national parks?
Cacheu River National Park, João Vieira Marine Park, and Orango Islands National Park
1234. What are administrative divisions called in Guinea-Bissau?
Regions
1235. How many regions does Guinea-Bissau have?
9
1236. What is the climate of Guinea-Bissau?
Tropical; generally hot and humid; monsoonal-type rainy season (June to November) with southwesterly winds; dry season (December to May) with northeasterly harmattan winds
1237. What are the natural resources of Guinea-Bissau?
Fish, timber, phosphates, bauxite, clay, granite, limestone, unexploited deposits of petroleum
1238. What are the natural hazards of Guinea-Bissau?
Hot, dry, dusty harmattan haze may reduce visibility during dry season; brush fires
1239. What are the religions of Guinea-Bissau?
Muslim 50%, indigenous beliefs 40%, Christian 10%
1240. What are the ethnic groups of Guinea-Bissau?
African 99% (includes Balanta 30%, Fula 20%, Manjaca 14%, Mandinga 13%, Papel 7%), European and mulatto less than 1%

Kenya

1241. What is the official name of Kenya?
Republic of Kenya

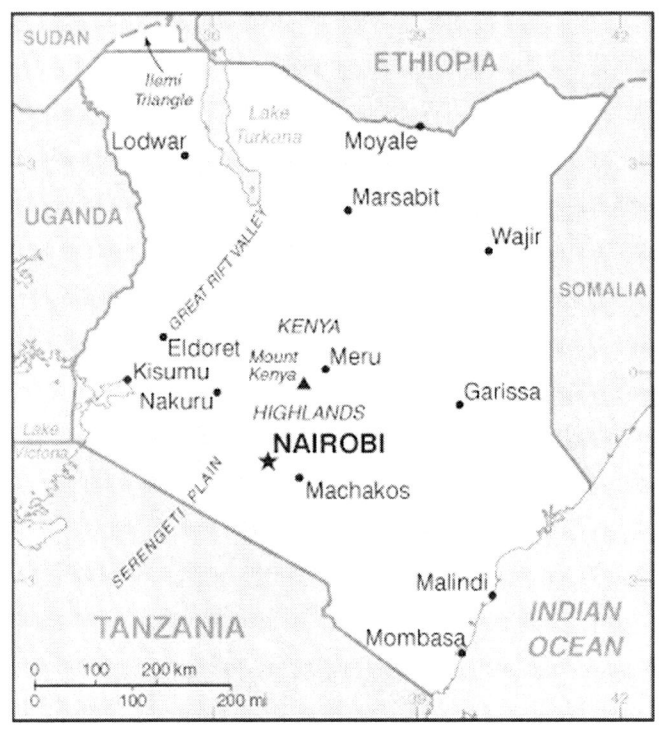

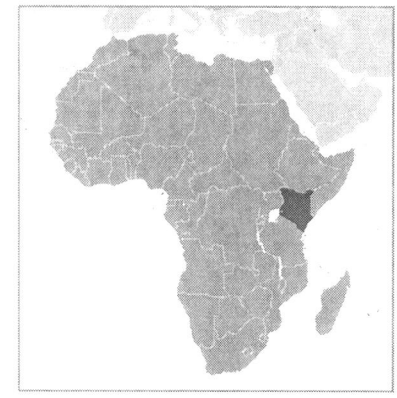

1242. Which country borders Kenya to the northwest?
Sudan
1243. Which country borders Kenya to the east?
Somalia
1244. Which country borders Kenya to the north?
Ethiopia
1245. Which country borders Kenya to the west?
Uganda
1246. Which country borders Kenya to the south?
Tanzania
1247. Which body of water borders Kenya to the southeast?
Indian Ocean
1248. Which body of water borders Kenya to the west?
Lake Victoria
1249. What is the motto of Kenya?
Let us all pull together
1250. What is the national anthem of Kenya?
O God of All Creation
1251. What is the capital of Kenya?
Nairobi (the largest city)
1252. What are official languages of Kenya?
Kiswahili and English
1253. What kind of government does Kenya have?
Semi-presidential Republic

1254. What was British East Africa?
A protectorate covering roughly the present-day Kenya from the 1880s to 1920
1255. What was established when the former East Africa Protectorate was transformed into a British crown colony in 1920?
The Colony of Kenya
1256. When did Kenya gain independence from United Kingdom?
December 12th, 1963
1257. When was the republic declared?
December 12th, 1964
1258. What is the area of Kenya?
224,080 sq mi / 580,367 km^2
1259. How long is the coastline of Kenya?
333 mi / 536 km
1260. What is the population of Kenya?
39,002,772 (by 2009)
1261. What is the currency of Kenya?
Kenyan Shilling
1262. What is the geographical feature of the terrain of Kenya?
Low plains rise to central highlands bisected by Great Rift Valley; fertile plateau in west
1263. What is the highest point of Kenya?
Mount Kenya (17,057 ft / 5,199 m)
1264. The country is named after what?
Mount Kenya
1265. What is Mount Kenya's rank by height in Africa?
2nd
1266. What is the peak of Mount Kenya?
Batian
1267. What type of mountain is Mount Kenya?
Stratovolcano
1268. Although Mount Kenya lies just south of the equator, it is still covered by ice. How many glaciers are there on Mount Kenya?
11
1269. How large is Mount Kenya National Park?
276 sq mi / 715 km^2
1270. For the surrounding region of Mount Kenya National Park, volcanic sediment in soil and the huge volume of fresh water coming down the slopes make it particularly favorable for what?
Agriculture
1271. What is the lowest point of Kenya?
Indian Ocean (0 ft / 0 m)
1272. Lake Turkana (also called Lake Rudolf) in Kenya extends its northern end into which country?
Ethiopia
1273. Lake Turkana is located in which valley?
Great Rift Valley

1274. What is Lake Turkana's rank by size among the world's permanent desert lake?
1st
1275. What is Lake Turkana's rank by size among the world's alkaline lake?
1st
1276. What is Lake Turkana's rank by size among the world's salt lake?
4th
1277. What is Central Island (also called Crocodile Island)?
A volcanic island located in the middle of Lake Turkana
1278. Which national park is composed of Central Island?
Central Island National Park
1279. What is inside of Central Island National Park?
More than a dozen craters and cones, three of which are filled by small lakes
1280. Which national park lies on the northeastern shore of Lake Turkana?
Sibiloi National Park
How large is the Sibiloi National Park, which is known for fossils?
607 sq mi / 1,571 km^2
1281. The Aberdare Range (also called Sattima Range) is a 99 mi (160 km) long mountain range in west central Kenya. It forms a section of the eastern rim of which valley?
Great Rift Valley
1282. What is the highest peak in the Aberdare Range?
Oldoinyo la Satima (13,120 ft /4,001 m)
1283. Which national park covers the higher areas of the Aberdare Mountain Range?
Aberdare National Park
1284. How many bird species are there in the Aberdare National Park?
250
1285. What animals live in the Aberdare National Park?
Lions, leopards, baboons, black and white Colobus monkeys, sykes monkeys, golden cats, melanistic serval cats, bongos, elands, and black rhino
1286. How large is the Amboseli National Park (also called Maasai Amboseli Game Reserve) spreaded across the Kenya-Tanzania border?
151 sq mi / 392 km^2
1287. What are the attractions of the Amboseli National Park?
Getting close to free-ranging elephants, meeting local people Maasai, and spectacular views of Mount Kilimanjaro
1288. How large is the Arabuko-Sokoke National Park, a small portion of the Arabuko-Sokoke Forest Reserve located on the coast of Kenya?
2 sq mi / 6 km^2
1289. What species are endemic or near endemic species in the Arabuko-Sokoke National Park?
Clarke's Weaver, Sokoke Scops Owl, Sokoke Pipit, Amani Sunbird and Spotted Ground Thrush, and Golden-rumped Elephant Shrew
1290. What kind of lake is Lake Naivasha, part of the Great Rift Valley?
Freshwater
1291. How large is Lake Naivasha?
54 sq mi / 139 km^2
1292. There is a sizeable population of what animals in Lake Naivasha?

Hippopotamuses
1293. Njorowa Gorge used to form the lake's outlet, but it is now high above the lake and forms the entrance to which national park?
Hell's Gate National Park
1294. How large is Hell's Gate National Park, which is famous for Fischer's Tower, Central Tower columns, and Hell's Gate Gorge?
26 sq mi / 68 km²
1295. Hell's Gate National Park has a wide variety of wildlife. Do they exist in large quantity?
No, many are few in number
1296. Mount Longonot is a dormant stratovolcano located southeast of which lake?
Lake Naivasha
1297. What is the elevation of Mount Longonot?
9,108 ft / 2,776 m
1298. Mount Longonot is home to what wildlife?
Zebra, giraffe, African buffalo, hartebeest, and Leopard
1299. Mount Longonot is part of which national part?
Mount Longonot National Park (690 sq mi / 1,788 km²)
1300. Kora National Park has what animal species?
Caracal, cheetah, elephant, genet, hippopotamus, spotted and striped hyenas, leopard, lion, serval, wildcat and several types of antelope
1301. What type of lake is Lake Nakuru in central Kenya?
Alkaline
1302. Lake Nakuru is famous for which bird specie?
Flamingo
1303. How large is Lake Nakuru National Park?
73 sq mi / 188 km²
1304. What kind of mountain is Mount Marsabit, located in northern Kenya?
Shield volcano
1305. How large is Marsabit National Park, located at Mount Marsabit?
600 sq mi / 1,554 km²
1306. Marsabit National Park is home to what bird species?
Sociable Weaver bird, Sparrow Weaver, and white-bellied turaco
1307. How large is Meru National Park, located in central Kenya?
336 sq mi / 870 km²
1308. What are attractions of Meru National Park?
The home of George and Joy Adamson, Adamson's Falls, views of Mount Kenya, and Tana River
1309. What is the longest river in Kenya?
Tana River (440mi / 800 km)
1310. Tana River originates in the Aberdare mountains and eventually turns into which two reservoirs by the Kindaruma dam?
Masinga Reservoir and Kiambere Reservoir
1311. Tana Rive enters the Indian Ocean at where?
Ungwana Bay
1312. Mount Elgon National Park is located on the border of Kenya and which country?

Uganda

1313. What are attractions of Mount Elgon National Park (493 sq mi / 1279 km²)?
Cliffs, caves, waterfalls, gorges, mesas, calderas, hot springs, and the mountain peaks

1314. Why are the four explorable caves in Mount Elgon National Park very popular?
Because there are frequent night visitors such as elephants and buffaloes that come to lick the natural salt found on the cave walls

1315. How high is the highest peak of Mount Elgon, Koitoboss, located on the Kenya side?
13,852 ft (4,155 m)

1316. What is Kenya's first national park?
Nairobi National Park (42 sq mi / 117 km²)

1317. Nairobi National Park is electrically fenced around the northern, eastern, and western boundaries. Its southern boundary is formed by which river?
The Mbagathi River

1318. What wildlife species are found in Nairobi National Park?
African buffalo, baboon, black rhinoceros, Burchell's zebra, cheetah, Coke's hartebeest, Grant's gazelle, hippopotamus, leopard, lion, Thomson's gazelle, eland, impala, Masai giraffe, ostrich, vulture, and waterbuck

1319. Fourteen Falls on the Athi River is close to which national park?
Ol Doinyo Sapuk National Park

1320. What is Kenya's smallest national park?
Saiwa Swamp National Park (1 sq mi / 3 km²)

1321. Inside Tsavo East National Park (4,536 sq mi / 11,747 km²), the Athi and Tsavo rivers converge to form which river?
Galana River

1322. What is the second longest river in Kenya?
Athi-Galana-Sabaki River (242 mi / 390 km)

1323. The Athi-Galana-Sabaki River starts as Athi River and changes to what name as it enters the Indian Ocean?
Galana River (also called Sabaki River)

1324. What is the Yatta Plateau inside Tsavo East National Park?
The world's longest lava flow, formed by lava from Ol Doinyo Sabuk Mountain

1325. How long is the lava flow of the Yatta Plateau?
180 mi / 290 km

1326. What is Lugard Falls inside Tsavo East National Park?
A series of rapids on the Galana River

1327. What is the Aruba Dam inside Tsavo East National Park?
A dam built in 1952 across the Voi River

1328. What is Tsavo West National Park (3500 sq mi / 9,065 km²) famous for?
Magnificent scenery, rich and varied wildlife (e.g. rhinoceros reserve), good road system, rock climbing and guided walks along the Tsavo River

1329. What divides Tsavo West National Park into east and west?
The A109 road between Nairobi and Mombasa, and the Uganda Railway

1330. The Uganda Railway is a historical railway system linking the interiors of Uganda and Kenya to the Indian Ocean at Mombasa in Kenya. What is the other name of the railroad?
Lunatic Express

1331. Kisite-Mpunguti Marine National Park (15 sq mi / 39 km^2) covers an area with how many small islands surrounded by coral reef?
    4
1332. What is the most heavily visited Kenyan marine park?
    Mombasa Marine National Park and Reserve
1333. What is special of Watamu Marine National Park and Reserve (81 sq mi / 210 km²)?
    Its coral gardens are 984 ft (300 m) from the shore and are home to approximately 600 species of fish, 110 species of stony coral and countless invertebrates, crustaceans and mollusks
1334. Masai Mara National Reserve is a large game reserve in south-western Kenya, which is the northern continuation of Serengeti National Park game reserve in which country?
    Tanzania
1335. What is Masai Mara National Reserve (583 sq mi / 1,510 km²) famous for?
    The exceptional population of big cats, game, and the annual migration of zebra, Thomson's gazelle and wildebeest from the Serengeti National Park from July to October
1336. What is the Great Migration?
    The migration in Masai Mara National Reserve is so immense that it is called the Great Migration
1337. What is the Kakamega Forest?
    The forest in western, near to the border with Uganda, with the area of 89 sq mi / 230 km²
1338. Kakamega Forest is said to be Kenya's last remnant of which ancient rainforest that once spanned the continent?
    Guineo-Congolian rainforest
1339. What is the largest indigenous montane forest in East Africa?
    Mau Forest in Kenya (1,055 sq mi / 2,732 km²)
1340. Which river is very important to supply water from Mount Kenya to the northern part of Kenya?
    Ewaso Ng'iro River
1341. What are administrative divisions called in Kenya?
    Provinces
1342. How many provinces does Kenya have?
    7 (plus 1 area)
1343. What is the climate of Kenya?
    Varies from tropical along coast to arid in interior
1344. What are the natural resources of Kenya?
    Limestone, soda ash, salt, gemstones, fluorspar, zinc, diatomite, gypsum, wildlife, hydropower
1345. What are the natural hazards of Kenya?
    Recurring drought; flooding during rainy seasons
1346. What are the religions of Kenya?
    Protestant 45%, Roman Catholic 33%, Muslim 10%, indigenous beliefs 10%, other 2%
1347. What are the ethnic groups of Kenya?
    Kikuyu 22%, Luhya 14%, Luo 13%, Kalenjin 12%, Kamba 11%, Kisii 6%, Meru 6%, other African 15%, non-African (Asian, European, and Arab) 1%

Lesotho

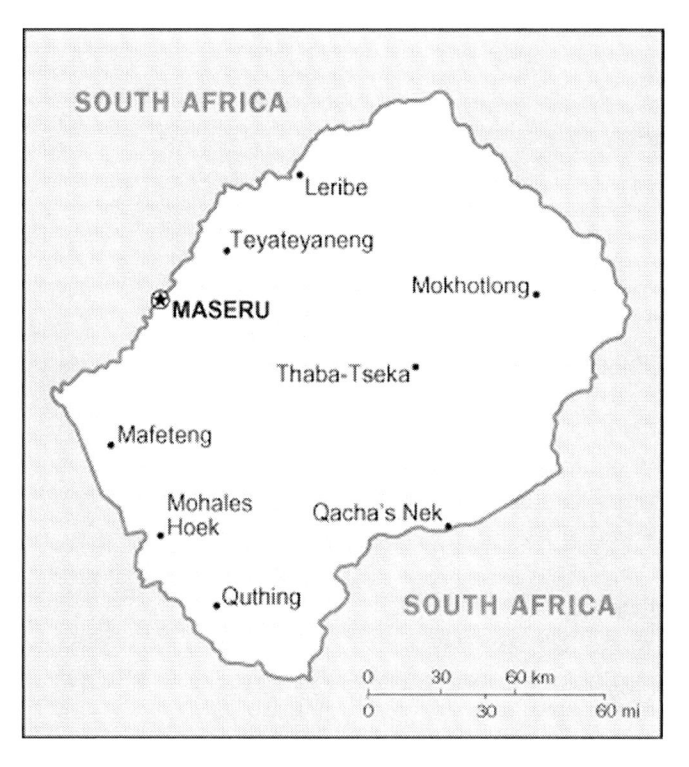

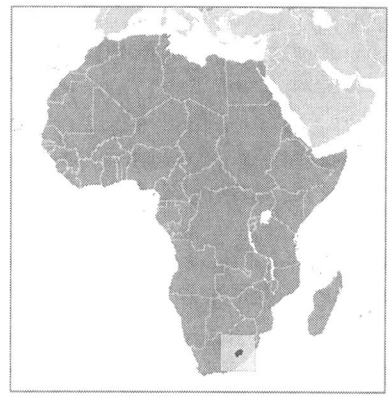

1348. What is the official name of Lesotho?
    Kingdom of Lesotho
1349. What is the meaning of "Lesotho"?
    The land of the people who speak Sesotho
1350. Lesotho is completely surrounded by which country?
    South Africa
1351. What is the significance of Lesotho's geographical position?
    It is the southernmost landlocked country in the world
1352. What is the motto of Lesotho?
    Peace, Rain, Prosperity
1353. What is the national anthem of Lesotho?
    Lesotho, land of our Fathers
1354. What is the capital of Lesotho?
    Maseru (the largest city)
1355. What are official languages of Lesotho?
    Sesotho and English
1356. What kind of government does Lesotho have?
    Parliamentary system and constitutional monarchy
1357. What was Basutoland?
    A British crown colony at present-day Lesotho during 1884–1966
1358. What was the official name of Basutoland?

Territory of Basutoland
1359. When did Lesotho gain independence from United Kingdom?
October 4th, 1966
1360. What is the area of Lesotho?
12,727 sq mi / 30,355 km²
1361. How long is the coastline of Lesotho?
0 mi / 0 km
1362. What is the population of Lesotho?
2,130,819 (by 2009)
1363. What is the currency of Lesotho?
Loti
1364. What is the geographical feature of the terrain of Lesotho?
Mostly highland with plateaus, hills, and mountains
1365. What is the highest point of Lesotho?
Thabana Ntlenyana (11,424 ft / 3,482 m)
1366. Thabana Ntlenyana is located in which mountain range?
Drakensberg (also called Maloti)
1367. What is Drakensberg's rank by height among the mountain ranges in Southern Africa?
1st
1368. Which four rivers rise in Drakensberg Mountains?
Tugela River, Orange River, Vaal River, and Caledon River
1369. Which river flows across the entire length of the country before exiting to South Africa?
Orange River (1,367 mi / 2,200 km)
1370. Which river, marking the northwestern part of the border with South Africa, is the primary source of water for Maseru?
Caledon River
1371. Which dam is on the Malibamat'so River in northern Lesotho?
Katse Dam
1372. Katse Dam, Africa's highest dam (at 185 m) and Africa's second largest dam, is part of which project in Lesotho?
Lesotho Highlands Water Project
1373. What is the lowest point of Lesotho?
Junction of the Orange and Makhaleng Rivers (4,593 ft / 1,400 m)
1374. Why is Lesotho the highest country in the world?
Because its lowest point is the highest in the world
1375. How large is Sehlabathebe National Park, located in the Drakensberg Mountains in Qacha's Nek District of Lesotho?
1,996 sq mi / 5,170 km²
1376. How large is Ts'ehlanyane National Park, which is home to several areas of ecological importance, which is home to the indigenous Leucosidea sericea woodland known locally as Ouhout or Che-che?
22 sq mi / 56 km²
1377. What are administrative divisions called in Lesotho?
Districts
1378. How many districts does Lesotho have?

1379. What is the climate of Lesotho?
Temperate; cool to cold, dry winters; hot, wet summers
1380. What are the natural resources of Lesotho?
Water, agricultural and grazing land, diamonds, sand, clay, building stone
1381. What are the natural hazards of Lesotho?
Periodic droughts
1382. What are the religions of Lesotho?
Christian 80%, indigenous beliefs 20%
1383. What are the ethnic groups of Lesotho?
Sotho 99.7%, Europeans, Asians, and other 0.3%

Liberia

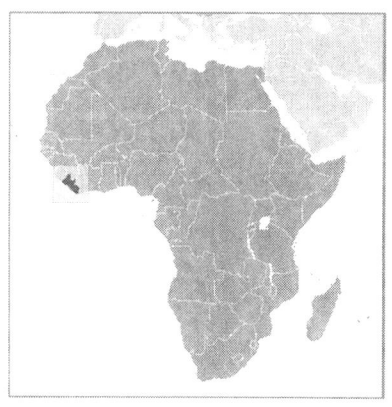

1384. What is the official name of Liberia?
Republic of Liberia
1385. Which country borders Liberia to the northwest?
Sierra Leone
1386. Which country borders Liberia to the north?
Guinea
1387. Which country borders Liberia to the east?
Côte d'Ivoire
1388. Which water body borders Liberia to the southwest?
Atlantic Ocean

1389. What is the motto of Liberia?
    The love of liberty brought us here
1390. What is the national anthem of Liberia?
    All Hail, Liberia, Hail!
1391. What is the capital of Liberia?
    Monrovia (the largest city)
1392. What is the official language of Liberia?
    English
1393. What kind of government does Liberia have?
    Republic
1394. How was Liberia founded?
    It was founded and colonized by freed American slaves in 1821–1822
1395. Which organization founded Liberia?
    The American Colonization Society
1396. When did Liberia gain independence from United States?
    July 26th, 1847
1397. What significant event happened in Liberia during 1980?
    A coup led by Samuel Doe which led to the First Liberian Civil War
1398. During which years did the First Liberian Civil War happen?
    1989–1996
1399. During which years did the Second Liberian Civil War happen?
    1999–2003
1400. What is the area of Liberia?
    43,000 sq mi / 111,369 km^2
1401. How long is the coastline of Liberia?
    360 mi / 579 km
1402. What is the population of Liberia?
    3,441,790 (by 2009)
1403. What is the currency of Liberia?
    Liberian Dollar
1404. What is the geographical feature of the terrain of Liberia?
    Mostly flat to rolling coastal plains rising to rolling plateau and low mountains in northeast
1405. What is the highest point of Liberia?
    Mount Wuteve (4,528 ft / 1,380 m, Data from the Shuttle Radar Topography Mission shows the height is 4,724 ft / 1,440 m)
1406. Mount Wuteve is located in which mountain range?
    Guinea Highlands
1407. Which mountain is taller than Mount Wuteve but is not wholly within Liberia?
    Mount Nimba (5,748 ft / 1,752 m)
1408. What is the lowest point of Liberia?
    Atlantic Ocean (0 ft / 0 m)
1409. What is Cape Mesurado?
    A headland on the coast of Liberia near the capital Monrovia and the mouth of the Saint Paul River
1410. The Saint Paul River forms part of the boundary between Liberia and which country?

1410. ... Guinea

1411. What is Cape Palmas?
A headland on the extreme southeast end of the coast of Liberia

1412. Cape Palmas marks the western limit of which gulf?
Gulf of Guinea

1413. What is Pepper Coast (also called Grain Coast)?
A coastal area in Liberia between Cape Mesurado and Cape Palmas

1414. What is the longest river in Liberia?
Cavalla River (also called Cavally, Youbou or Diougou River, 320 mi / 515 km)

1415. The Cavalla River forms the southeast boundary of Liberia with which country?
Côte d'Ivoire

1416. The Mano River originates in the Guinea Highlands in Liberia and forms part of the northwestern Liberia with which country?
Sierra Leone

1417. How long is the Saint John River, with its headwaters in neighboring Guinea, flowing generally southwest through Liberia and emptying into the Atlantic Ocean at Bassa Cove?
175 mi (282 km)

1418. The Cestos River rises in the Nimba Range of Guinea, flows south along the Côte d'Ivoire border, then southwest through tracks of Liberian rain forest to empty into where?
Atlantic Ocean

1419. How many national parks does Liberia have?
1 (Sapo National Park, 697 sq mi / 1,804 km^2)

1420. What are administrative divisions called in Liberia?
Counties

1421. How many counties does Liberia have?
15

1422. What is the climate of Liberia?
Tropical; hot, humid; dry winters with hot days and cool to cold nights; wet, cloudy summers with frequent heavy showers

1423. What are the natural resources of Liberia?
Iron ore, timber, diamonds, gold, and hydropower

1424. What are the natural hazards of Liberia?
Dust-laden harmattan winds blow from the Sahara (December to March)

1425. What are the religions of Liberia?
Christian 40%, Muslim 20%, indigenous beliefs 40%

1426. What are the ethnic groups of Liberia?
Indigenous African 95% (including Kpelle, Bassa, Gio, Kru, Grebo, Mano, Krahn, Gola, Gbandi, Loma, Kissi, Vai, Dei, Bella, Mandingo, and Mende), Americo-Liberians 2.5% (descendants of immigrants from the US who had been slaves), Congo People 2.5% (descendants of immigrants from the Caribbean who had been slaves)

Libya

1427. What is the official name of Libya?
Great Socialist People's Libyan Arab Jamahiriya

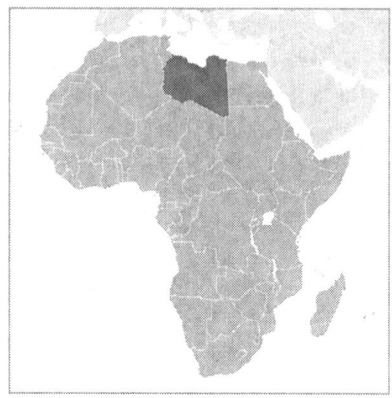

1428. Which country borders Libya to the southeast?
    Sudan
1429. Which country borders Libya to the east?
    Egypt
1430. Which countries border Libya to the south?
    Chad and Niger
1431. Which country borders Libya to the west?
    Algeria
1432. Which country borders Libya to the northwest?
    Tunisia
1433. Which water body borders Libya to the north?
    Mediterranean Sea
1434. What is the motto of Libya?
    Liberty, Socialism, Unity
1435. What is the national anthem of Libya?
    God is the Greatest
1436. What is special about the flag of Libya?
    It consists of a green field with no other characteristics, and it is the only national flag in the world with just one color and no design, insignia, or other details
1437. What is the capital of Libya?
    Tripoli (the largest city)
1438. What is the official language of Libya?
    Arabic
1439. What kind of government does Libya have?

92

Jamahiriya (Arabic term of Republic)
1440. Which three historic regions was Libya divided into?
Cyrenaica, Tripolitania, and Fezzan
1441. What were these regions called in the pre-1963 administrative system?
Province or state
1442. Which historic region was the eastern coastal region of Libya?
Cyrenaica
1443. Which historic region included Tripoli?
Tripolitania
1444. Which historic region was largely desert in the southwestern Libya?
Fezzan
1445. Is it true that Egyptians, Greeks, Vikings, Romans, and Byzantines were present in Libya before 642 CE?
No, not Vikings
1446. Islamic rule was in existence in which two regions during 642–1551?
Tripolitania and Cyrenaica
1447. Who ruled Libya during 1551–1912?
Ottoman
1448. When did Ottoman Libya become Italian colony under Treaty of Lausanne?
1912
1449. Italian forces were driven out by Allies of World War II in which year?
1943
1450. When was Libya relinquished by Italy?
February 10th, 1947
1451. Who ruled Libya after Italy?
 United Kingdom and France under United Nations Trusteeship (The United Kingdom was responsible for Tripolitania and Cyrenaica and France was responsible for Fezzan)
1452. When did Libya gain independence from United Kingdom and France under United Nations Trusteeship?
December 24th, 1951
1453. Kingdom of Libya came into existence on December 24, 1951 and lasted until when?
September 1st, 1969
1454. A 1969 military coup by Qadhafi turned the country into what kind of government?
Combination of socialism and theocracy
1455. When did United States rescind Libya's designation as a state sponsor of terrorism?
June 2006
1456. When did United States and Libya exchange ambassadors for the first time since 1973?
January 2009
1457. What is the area of Libya?
679,359 sq mi / 1,759,541 km^2
1458. What is Libya's rank by area in Africa?
4th
1459. What is Libya's rank by area in world?
7th
1460. How long is the coastline of Libya?

1,100 mi / 1,770 km
1461. What is the population of Libya?
6,310,434 (by 2009)
1462. What is the currency of Libya?
Dinar
1463. What is the geographical feature of the terrain of Libya?
Mostly barren, flat to undulating plains, plateaus, depressions
1464. What percentage of Libya is desert?
90%
1465. What is the highest point of Libya?
Bikku Bitti (7,438 ft / 2,267 m)
1466. Bikku Bitti is located in which mountain range?
Tibesti Mountains
1467. The Tibesti Mountains are a volcanic group of inactive volcanoes located in southern Libya and which other country?
Chad
1468. What is the lowest point of Libya?
Sabkhat Ghuzayyil (-154 ft / -47 m)
1469. What is Gulf of Sidra (also called Gulf of Sirte)?
A water body in the Mediterranean Sea on the northern coast of Libya
1470. The portion of the Mediterranean Sea north of Libya is often called what?
Libyan Sea
1471. Which ethnic group is concentrated in Nafusa Mountains, a region in western Libya?
Berbers
1472. The Libyan Desert, located in the northern and eastern part of the Sahara Desert, occupies Libya and which other two countries?
Egypt and Sudan
1473. The Libyan Desert includes which three sand seas, which contain dunes up to 361 ft (110 m) in height and cover about 25% of the Libyan Desert?
Rebiana Sand Sea, Egyptian Sand Sea, and Kalansho Sand Sea
1474. What are administrative divisions called in Libya?
Municipalities
1475. How many municipalities does Libya have?
25
1476. What is the climate of Libya?
Mediterranean along coast; dry, extreme desert interior
1477. What are the natural resources of Libya?
Petroleum, natural gas, gypsum
1478. What are the natural hazards of Libya?
Hot, dry, dust-laden ghibli is a southern wind lasting one to four days in spring and fall; dust storms, sandstorms
1479. What are the religions of Libya?
Sunni Muslim 97%, other 3%
1480. What are the ethnic groups of Libya?

Berber and Arab 97%, other 3% (includes Greeks, Maltese, Italians, Egyptians, Pakistanis, Turks, Indians, and Tunisians)

Madagascar

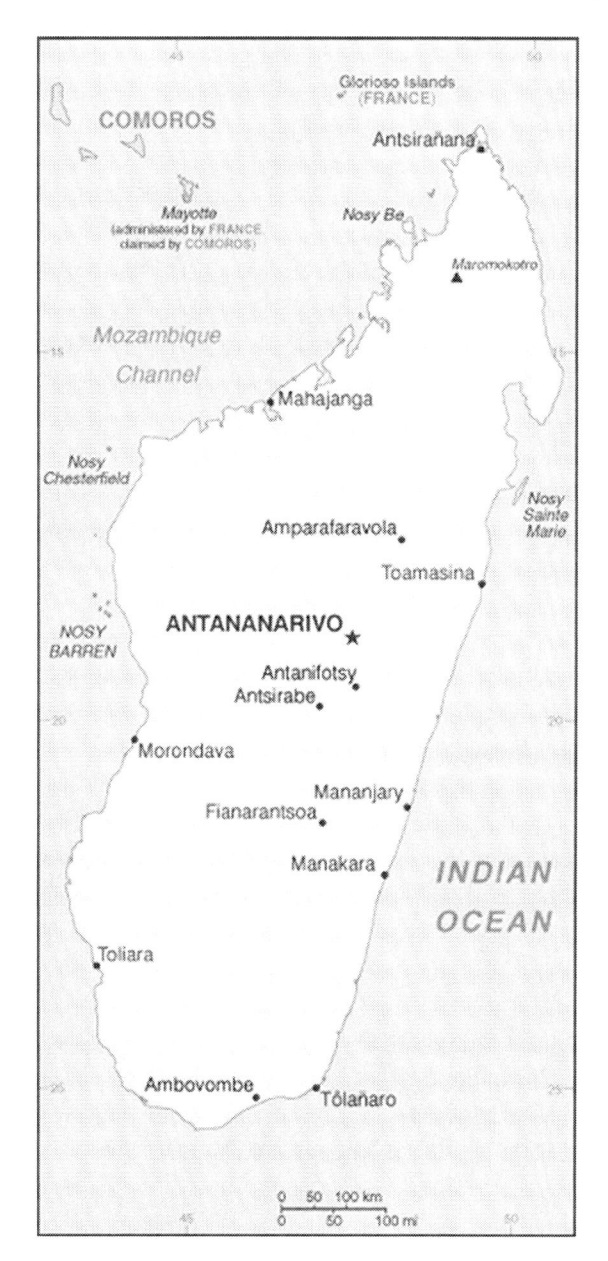

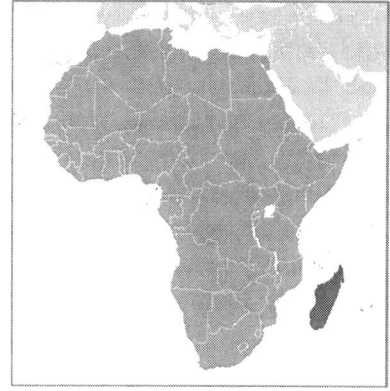

1481. What is the official name of Madagascar?
    Republic of Madagascar
1482. Madagascar is an island nation in which water body?
    Indian Ocean

1483. The Mozambique Channel is a portion of the Indian Ocean located between Madagascar and which country?
Mozambique
1484. Madagascar is home to 5% of the world's plant and animal species, of which what percentage are endemic to Madagascar
More than 80%
1485. What is the motto of Madagascar?
Fatherland, Liberty, Progress
1486. What is the national anthem of Madagascar?
Oh, Our Beloved Fatherland
1487. What is the capital of Madagascar?
Antananarivo (the largest city)
1488. What are official languages of Madagascar?
Malagasy, French, and English
1489. What kind of government does Madagascar have?
Caretaker government
1490. After a brief de facto protectorate period beginning in 1885, Madagascar became a full formal French protectorate in which year?
1890
1491. When did Madagascar become a French colony?
1896
1492. When did Madagascar gain independence from France?
June 26th, 1960
1493. What is the area of Madagascar?
226,597 sq mi / 587,041 km^2
1494. What is the Madagascar Island's rank by size in the world?
4th
1495. How long is the coastline of Madagascar?
3,000 mi / 4,828 km
1496. What is the population of Madagascar?
20,653,556 (by 2009)
1497. What is the currency of Madagascar?
Malagasy Ariary
1498. What is the geographical feature of the terrain of Madagascar?
Narrow coastal plain, high plateau and mountains in center
1499. Madagascar is divided into which five geographical regions?
The east coast, the Tsaratanana Massif, the central highlands, the west coast, and the southwest
1500. What is the highest point of Madagascar?
Maromokotro (9,436 ft / 2,876 m)
1501. Which geographical region is Maromokotro located?
The Tsaratanana Massif
1502. What are two prominent features of the Tsaratanana Massif?
The excellent natural harbor at Antsiranana and the large island of Nosy-Be to the west
1503. What is the lowest point of Madagascar?

Indian Ocean (0 ft / 0 m)
1504. What is the Canal des Pangalanes in the east coast region?
A series of man made and natural lakes linked by rivers running down the east coast of Madagascar for 286 miles (460 km)
1505. Which national park, located on Masoala Peninsula, is Madagascar's largest protected area?
Masoala National Park
1506. How large is the Masoala National Park?
927 sq mi / 2400 km^2 (888 sq mi / 2,300 km^2 of rainforest and 39 sq mi / 100 km^2 of marine parks)
1507. How many lemur species are there in the Masoala National Park?
10
1508. What are the three marine parks in the Masoala National Park?
Tampolo in the West, Ambodilaitry in the South, and Ifaho in the East
1509. The central highlands, which range from 2,625 ft to 5,906 ft (800 m to 1,800 m) in altitude, contain what kind of topographies?
Rounded and eroded hills, massive granite outcrops, extinct volcanoes, eroded peneplains, alluvial plains and marshes
1510. Antananarivo is located in which high plateau in the near centre of Madagascar?
Hauts Plateaux
1511. Deep bays and well-protected harbors in which region have attracted explorers, traders, and pirates from Europe, Africa, and the Middle East since ancient times?
The west coast
1512. The southwest is bordered on the east by the Ivakoany Massif and on the north by which massif?
The Isala Roiniforme Massif
1513. Which national park in the northern Madagascar is known for its water falls, crater lakes, different species of birds, 25 species of mammals, and 59 species of reptiles
Amber Mountain National Park
1514. How large is the Andohahela National Park located in southeastern Madagascar, known for the extremes of habitats?
293 sq mi / 760 km²
1515. How large is the Andringitra National Park in southeastern Madagascar, known for its rough terrain, over 100 different species of birds, over 50 species of mammals, and 55 species of frogs?
12,031 sq mi / 31,160 km²
1516. How large is the Ankarafantsika National Park, located in northwestern Madagascar, which is a mosaic of dense dry deciduous forest forests and wetlands?
50,203 sq mi / 130,026 km²
1517. How large is the Isalo National Park, located in Southwestern Madagascar, which is known for its wide variety of terrain (sandstone formations, deep canyons, palm-lined oases, and grassland) and Flora and Fauna (82 species of birds, 33 species of reptiles, 15 species of frogs and 14 species of mammals, 340 different species of fauna)?
315 sq mi / 815 km²

1518. How large is Andasibe-Mantadia National Park in eastern Madagascar, where the rainforest is habitat to a vast species biodiversity, including many endemic rare species and endangered species, including 11 lemur species?
60 sq mi / 155 km²
1519. What are he Andasibe-Mantadia National Park's two component parts?
Mantadia National Park and Analamazoatra Reserve
1520. How large is the Ranomafana National Park, located in southeastern Madagascar, which is home to several rare species of flora and fauna such as the lemur?
161 sq mi / 417 km²
1521. What is special of the Tsingy de Bemaraha Strict Nature Reserve, located close to the western coast of Madagascar?
Unique geography, preserved mangrove forests, and wild bird and lemur populations of the area
1522. How large is the Lokobe Reserve, located in northwestern Madagascar, which is known for its black lemurs and the beautiful Nosy Be Panther Chameleon?
6 sq mi / 15 km²
1523. How large is the Tsingy de Namoroka Strict Nature Reserve (also called Namoroka National Park) in the northwestern Madagascar, which is known for its tsingy walls, caves, canyons, and natural swimming pools?
85 sq mi / 221 km²
1524. How large is the Ankarana Reserve, a small vegetated plateau in northern Madagascar?
70 sq mi / 182 km²
1525. What is Madagascar's largest lemur?
Indri
1526. What is the largest lake in Madagascar?
Lake Alaotra (347 sq mi / 900 km²)
1527. What are administrative divisions called in Madagascar?
Provinces
1528. How many provinces does Madagascar have?
6
1529. What is the climate of Madagascar?
Tropical along coast, temperate inland, arid in south
1530. What are the natural resources of Madagascar?
Graphite, chromite, coal, bauxite, salt, quartz, tar sands, semiprecious stones, mica, fish, hydropower
1531. What are the natural hazards of Madagascar?
Periodic cyclones; drought; and locust infestation
1532. What are the religions of Madagascar?
Indigenous beliefs 52%, Christian 41%, Muslim 7%
1533. What are the ethnic groups of Madagascar?
Malayo-Indonesian (Merina and related Betsileo), Cotiers (mixed African, Malayo-Indonesian, and Arab ancestry - Betsimisaraka, Tsimihety, Antaisaka, Sakalava), French, Indian, Creole, Comoran

Malawi

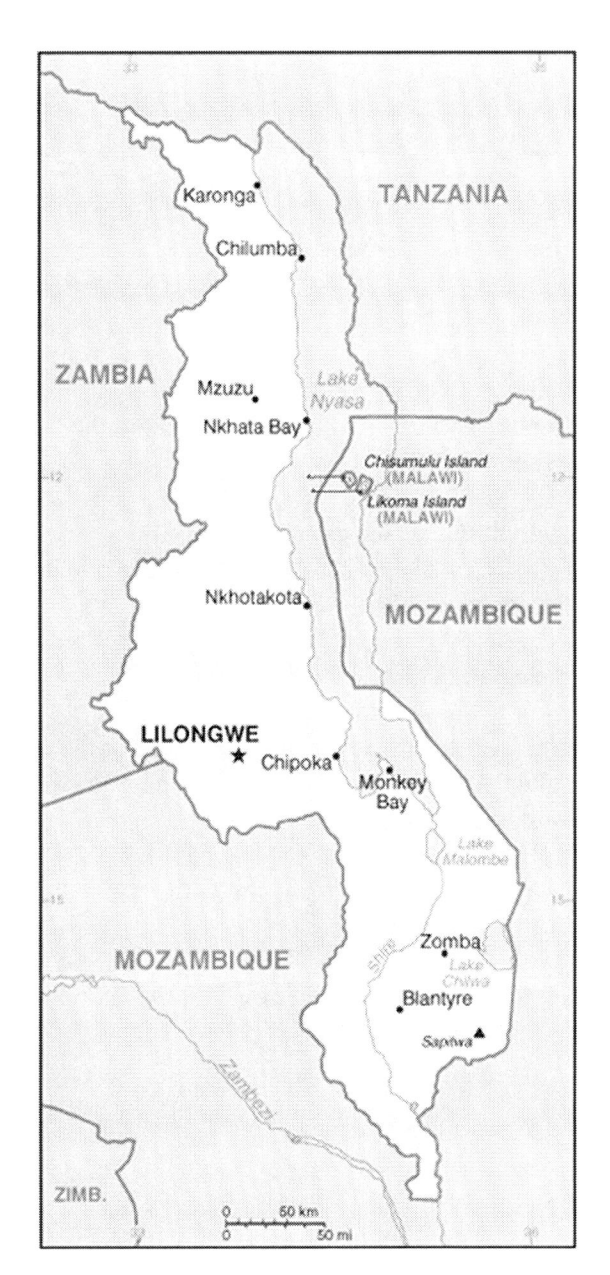

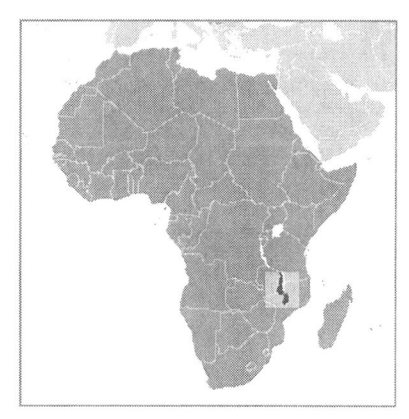

1534. What is the official name of Malawi?
    Republic of Malawi
1535. Which country borders Malawi to the east, south and west?
    Mozambique
1536. Which country borders Malawi to the northwest?
    Zambia
1537. Which country borders Malawi to the northeast?
    Tanzania
1538. What is the motto of Malawi?
    Unity and Freedom

1539. What is the national anthem of Malawi?
Oh God Bless Our Land of Malawi
1540. What is the capital of Malawi?
Lilongwe
1541. What is the largest city of Malawi?
Blantyre
1542. What is the official language of Malawi?
English
1543. What kind of government does Malawi have?
Multi-party democracy
1544. Which empire was a state established by Bantu people in the area of Lake Malawi, in the present-day Malawi, sometime during the 16th century?
Maravi Empire
1545. Malawi was the British Central Africa Protectorate during which period of time?
1893–1907
1546. What was the capital of the British Central Africa Protectorate?
Zomba
1547. When was British Central Africa Protectorate renamed to Nyasaland Protectorate?
July 6th, 1907
1548. What was Federation of Rhodesia and Nyasaland (also called Central African Federation)?
A federal realm of the British Crown — not a colony and not a dominion, with the intention to eventually become a dominion in the Commonwealth of Nations
1549. When did Federation of Rhodesia and Nyasaland exist?
August 1st, 1953–December 31st, 1963
1550. When did Malawi gain independence from United Kingdom?
July 6th, 1964
1551. When did Malawi vote to change one-party rule to multi-party democracy?
June 14th, 1993
1552. What is the area of Malawi?
45,747 sq mi / 118,484 km^2
1553. How long is the coastline of Malawi?
0 mi / 0 km
1554. What is the population of Malawi?
14,268,711 (by 2009)
1555. What is the currency of Malawi?
Kwacha
1556. What is the geographical feature of the terrain of Malawi?
Narrow elongated plateau with rolling plains, rounded hills, some mountains
1557. What is the highest point of Malawi?
Sapitwa (also called Mount Mlanje, 9,849 ft / 3,002 m)
1558. What is the lowest point of Malawi?
Junction of the Shire River and international boundary with Mozambique (121 ft / 37 m)
1559. Which lake separates Malawi from Tanzania and Mozambique?
Lake Malawi (also called Lake Nyasa, Lake Nyassa, Lake Niassa, or Lago Niassa)
1560. Why is Lake Malawi sometimes called the Calendar Lake?

Because it is about 365 miles (587 km) long and 52 miles (84 km) wide
1561. Is Lake Malawi an African Great Lake?
Yes
1562. What is Lake Malawi's rank by surface area in Africa?
3rd (11,429 sq mi / 29,600 km²)
1563. What is Lake Malawi's rank by surface area in the world?
8th
1564. What is Lake Malawi's rank by depth in Africa?
2nd
1565. What type of lake is Lake Malawi?
Rift, freshwater
1566. What is the primary inflow of Lake Malawi?
Ruhuhu River
1567. What is the primary outflow of Lake Malawi?
Shire River
1568. How long is the Shire River?
250 mi / 402 km (including Lake Malawi and Ruhuhu River, its headstream, Shire River has a length of about 746 mil / 1200 km)
1569. Where does the Shire River flow into?
Zambezi River
1570. The upper Shire River connects Lake Malawi with which lake?
Lake Malombe
1571. What is the surface area of Lake Malombe?
174 sq mi / 450 km²
1572. How long is the Dwangwa River, a tributary river for Lake Malawi?
100 mi / 160 km
1573. What is the second-largest lake in Malawi?
Lake Chilwa
1574. Why was Lake Malawi National Park inscribed as a UNESCO World Heritage Site in 1984?
Because many endemic fish species make it a key example of specialized evolution
1575. Which place is located in the Central Region of Malawi and was inscribed as a UNESCO World Heritage Site in 2006?
Chongoni Rock Art Area
1576. How large is Lake Malawi National Park in Malawi?
36 sq mi / 94 km²
1577. How large is the Kasungu National Park, located in the Central Region of Malawi, which is known for its population of elephants?
894 sq mi / 2,316 km²
1578. How large is the Lengwe National Park in Malawi, located in southern Malawi, which is home to the Nyala Antelope?
342 sq mi / 887 km²
1579. How large is the Liwonde National Park located on the upper Shire River plain?
212 sq mi / 548 km²
1580. How large is the Mwabvi Wildlife Reserve in southern Malawi, known for the last natural home to Malawi's Black Rhino population?

131 sq mi / 340 km²

1581. How large is the Vwaza Marsh Game Reserve Malawi, known for large herds of buffalo and elephants, hippopotamuses and antelopes?
386 sq mi / 1,000 km²

1582. What is the largest national park in Malawi?
Nyika National Park (1,250 sq mi / 3,200 km²)

1583. The Nyika National Park covers practically the whole of which plateau in northern Malawi?
Nyika Plateau

1584. Nyika National Park extends to which other country?
Zambia

1585. What are administrative divisions called in Malawi?
Districts

1586. How many districts does Malawi have?
28

1587. What is the climate of Malawi?
Sub-tropical; rainy season (November to May); dry season (May to November)

1588. What are the natural resources of Malawi?
Limestone, arable land, hydropower, unexploited deposits of uranium, coal, and bauxite

1589. What are the religions of Malawi?
Christian 79.9%, Muslim 12.8%, other 3%, none 4.3%

1590. What are the ethnic groups of Malawi?
Chewa, Nyanja, Tumbuka, Yao, Lomwe, Sena, Tonga, Ngoni, Ngonde, Asian, European

Mali

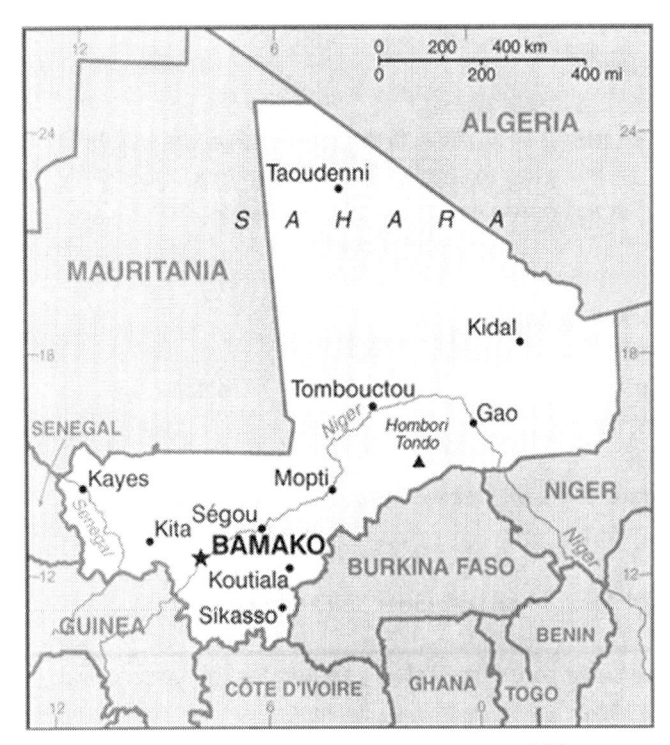

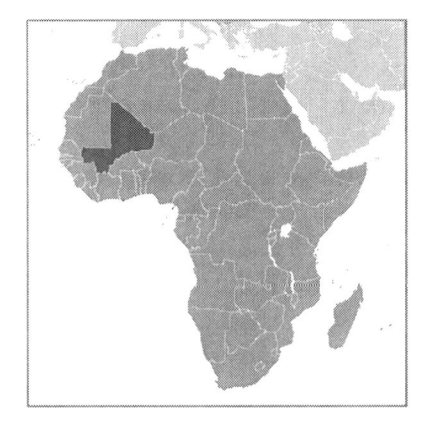

1591. What is the official name of Mali?
    Republic of Mali
1592. Which countries border Mali to the west?
    Senegal and Mauritania
1593. Which country borders Mali to the northeast?
    Algeria
1594. Which country borders Mali to the east?
    Niger
1595. Which country borders Mali to the southwest?
    Guinea
1596. Which countries border Mali to the south?
    Burkina Faso and Côte d'Ivoire
1597. What is the motto of Mali?
    One people, one goal, one faith
1598. What is the national anthem of Mali?
    Mali
1599. What is the capital of Mali?
    Bamako (the largest city)
1600. What is the official language of Mali?
    French
1601. What kind of government does Mali have?
    Semi-presidential republic
1602. Was Mali part of the Ghana Empire (also called Wagadou Empire) during 790–1076?
    Yes
1603. When was Mali Empire (also called Manding Empire, or Manden Kurufa) established?
    1230s
1604. What was the capital of Mali Empire?
    Niani
1605. When did the capital of Mali Empire move from Niani to Kangaba?
    1559
1606. When did Mali Empire collapse and be divided among emperor's sons?
    1600s
1607. What language was used in Mali Empire?
    Mandinkan
1608. What was the religion of Mali Empire?
    Ancestor Worship; Islam
1609. What was the currency used in Mali Empire?
    Gold Dust
1610. What was the population in Mali Empire?
    45,000,000 (by 1450)
1611. How large was Mali Empire by 1312?
    500,000 sq mi / 1,294,994 km^2
1612. The Songhai Empire was freed from which empire in 1340?
    Mali Empire
1613. What was the capital of Songhai Empire?

Gao
1614. What language was used in Songhai Empire?
Songhai
1615. What was the religion of Songhai Empire?
Islam
1616. What was the currency used in Songhai Empire?
Cowries
1617. How large was Songhai Empire by 1500?
540,543 sq mi / 1,400,000 km^2
1618. When was Songhai Empire destroyed?
1591
1619. The Songhai people established Dendi Kingdom in which present-day country after the collapse of Songhai Empire?
Niger
1620. French Sudan was created as a colony in French West Africa on September 9, 1880 as Upper Senegal, and was renamed to what on August 18, 1890?
French Sudan Territory
1621. What was the capital of French Sudan until 1899?
Kayes
1622. What was the capital of French Sudan from 1899 to its final collapse on 1959?
Bamako
1623. French Sudan was dissolved into Senegambia and Niger, an administrative unit of the French possessions during which time period?
1902–1904
1624. Senegambia and Niger was reorganized in 1904 into what?
Upper Senegal and Niger
1625. On October 10, 1899 French Sudan was broken up, and 11 southern provinces went to which countries?
French Guinea (present-day Guinea), the Côte d'Ivoire and Dahomey (present-day Benin)
1626. How large was French Sudan in 1959?
479,245 sq mi / 1,241,238 km^2
1627. When was French Sudan dissolved into Mali Federation?
April 4, 1959
1628. What was the capital of Mali Federation?
Dakar
1629. Mali Federation was a union between French Sudan and which country?
Senegal
1630. When did Mali Federation gain independence from France?
June 20th, 1960
1631. When did Mali Federation collapse, due to Senegal secession?
August 20th, 1960
1632. When did French Sudan proclaim itself the Republic of Mali and withdrew from the French Community?
September 22nd, 1960
1633. What is the area of Mali?

478,839 sq mi / 1,240,192 km²

1634. How long is the coastline of Mali?
0 mi / 0 km

1635. What is the population of Mali?
12,666,987 (by 2009)

1636. What is the currency of Mali?
West African CFA Franc

1637. What is the geographical feature of the terrain of Mali?
Mostly flat to rolling northern plains covered by sand; savanna in south, rugged hills in northeast

1638. Mali is divided into which three natural zones?
Southern cultivated Sudanese zone, central semiarid Sahelian zone, and northern arid Saharan zone

1639. What is the highest point of Mali?
Hombori Tondo (3,789 ft / 1,155 m)

1640. What is the lowest point of Mali?
Senegal River (75 ft / 23 m)

1641. Which river and Senegal River are Mali's two largest rivers?
Niger River

1642. Which river is generally described as Mali's lifeblood, a source of food, drinking water, irrigation, and transportation?
Niger River

1643. Which national park, located in southern Mali, is home for monkeys and some chimpanzees?
Bafing National Park

1644. Which national park, located in western Mali, is known for its prehistoric rock art and tombs?
Boucle du Baoulé National Park (1,286 sq mi / 3,330 km²)

1645. What are administrative divisions called in Mali?
Regions

1646. How many regions does Mali have?
8

1647. What is the climate of Mali?
Subtropical to arid; hot and dry (February to June); rainy, humid, and mild (June to November); cool and dry (November to February)

1648. What are the natural resources of Mali?
Gold, phosphates, kaolin, salt, limestone, uranium, gypsum, granite, hydropower

1649. What are the natural hazards of Mali?
Hot, dust-laden harmattan haze common during dry seasons; recurring droughts; occasional Niger River flooding

1650. What are the religions of Mali?
Muslim 90%, Christian 1%, indigenous beliefs 9%

1651. What are the ethnic groups of Mali?
Mande 50% (Bambara, Malinke, Soninke), Peul 17%, Voltaic 12%, Songhai 6%, Tuareg and Moor 10%, other 5%

Mauritania

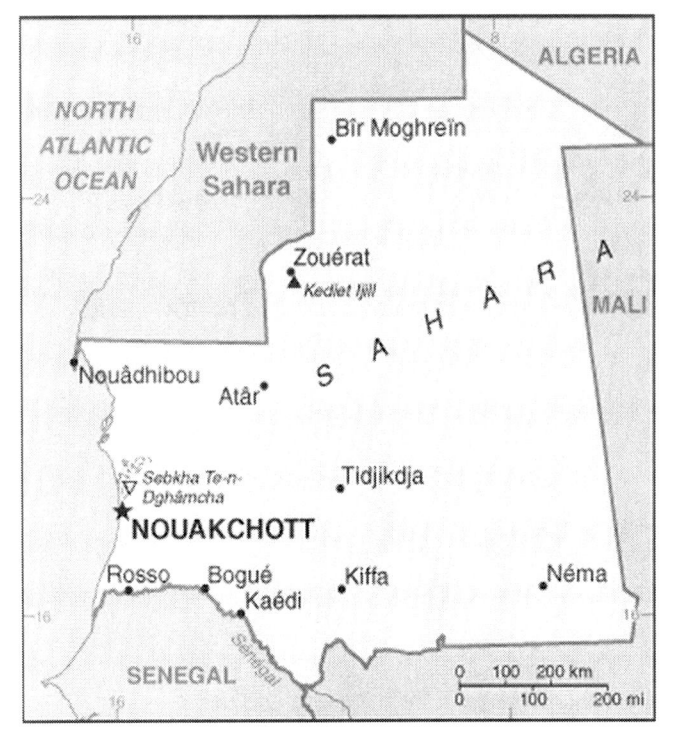

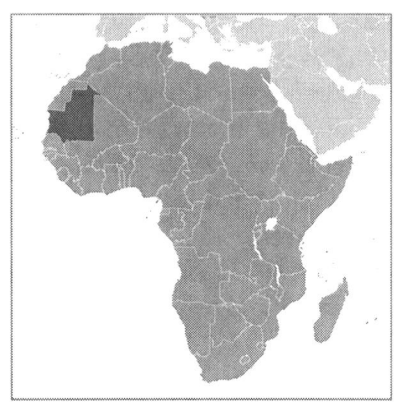

1652. What is the official name of Mauritania?
    Islamic Republic of Mauritania
1653. Which country borders Mauritania to the southwest?
    Senegal
1654. Which country borders Mauritania to the northeast?
    Algeria
1655. Which territory borders Mauritania to the northwest?
    Western Sahara
1656. Which country borders Mauritania to the east and southeast?
    Mali
1657. Which water body borders Mauritania to the west?
    Atlantic Ocean
1658. What is the motto of Mauritania?
    Honor, Fraternity, Justice
1659. What is the national anthem of Mauritania?
    National anthem of Mauritania
1660. What is the capital of Mauritania?
    Nouakchott (the largest city)
1661. What is the official language of Mauritania?
    Arabic
1662. What kind of government does Mauritania have?

Military junta
1663. The Kingdom of Waalo (also called Oualo) was a kingdom on the lower Senegal River in West Africa during which period of time?
1287–1855
1664. The Kingdom of Waalo included Mauritania and which country?
Senegal
1665. Which country conquered Mauritania in 1855?
France
1666. When did Mauritania gain independence from France?
November 28th, 1960
1667. What is the area of Mauritania?
397,954 sq mi / 1,030,700 km^2
1668. How long is the coastline of Mauritania?
469 mi / 754 km
1669. What is the population of Mauritania?
3,129,486 (by 2009)
1670. What is the currency of Mauritania?
Ouguiya
1671. What is the geographical feature of the terrain of Mauritania?
Mostly barren, flat plains of the Sahara; some central hills
1672. Mauritania is divided into which four ecological zones?
The Saharan Zone, the Sahelian Zone, the Senegal River Valley, and the Coastal Zone
1673. What is the highest point of Mauritania?
Kediet Ijill (3,002 ft / 915 m)
1674. What is the lowest point of Mauritania?
Sebkhet Te-n-Dghamcha (-16 ft / -5 m)
1675. What is the highest sandstone plateau in Mauritania?
Adrar Plateau (1,640 ft / 500 m)
1676. What is the Richat Structure in the Sahara desert of the north-central region in Mauritania?
A prominent circular feature with a diameter of almost 30 miles (50 km), forming a conspicuous bull's-eye in the otherwise rather featureless expanse of the desert
1677. How large is Banc d'Arguin National Park, located on the west coast of Mauritania, which is a major breeding site for migratory birds?
4,633 sq mi / 12,000 km²
1678. How large is Diawling National Park, located in southwestern Mauritania, which is known for over 220 species of birds and for its fish?
62 sq mi / 160 km²
1679. What are administrative divisions called in Mauritania?
Regions
1680. How many regions does Mauritania have?
12 (plus 1 capital district)
1681. What is the climate of Mauritania?
Desert; constantly hot, dry, dusty
1682. What are the natural resources of Mauritania?
Iron ore, gypsum, copper, phosphate, diamonds, gold, oil, and fish

1683. What are the natural hazards of Mauritania?
    Hot, dry, dust/sand-laden sirocco wind blows primarily in March and April; periodic droughts
1684. What are the religions of Mauritania?
    Muslim 100%
1685. What are the ethnic groups of Mauritania?
    Mixed Moor/black 40%, Moor 30%, black 30%

Mauritius

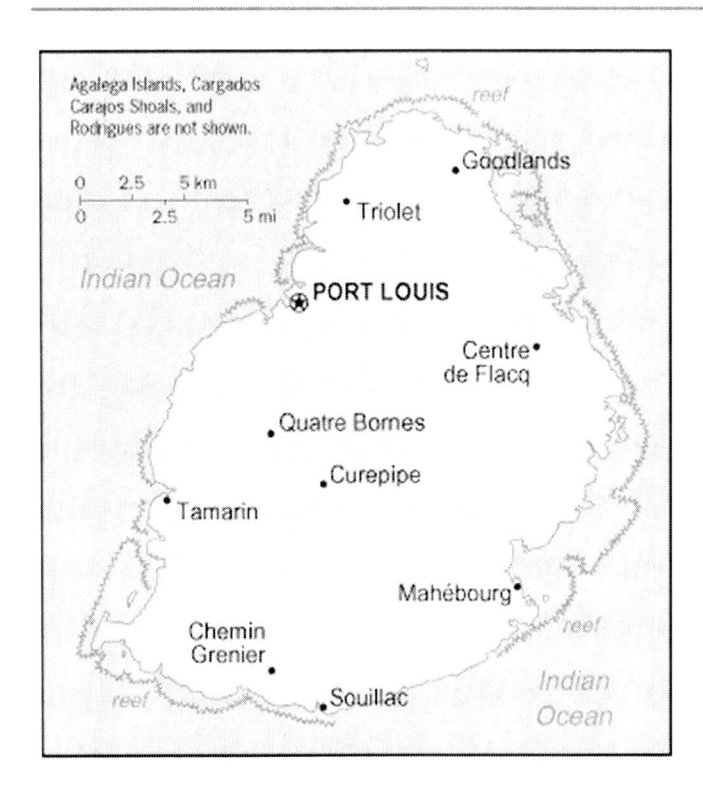

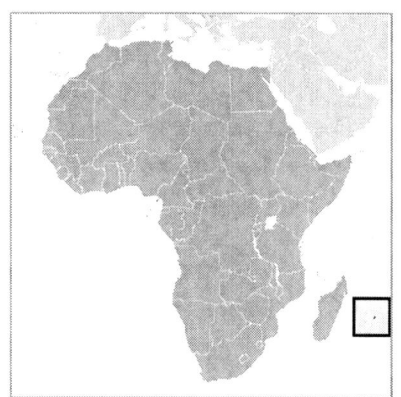

1686. What is the official name of Mauritius?
    Republic of Mauritius
1687. Mauritius is about 560 miles (900 km) east of which country?
    Madagascar
1688. What is the motto of Mauritius?
    Star and Key of the Indian Ocean
1689. What is the national anthem of Mauritius?
    Motherland
1690. What is the capital of Mauritius?
    Port Louis (the largest city)
1691. What is the official language of Mauritius?
    English
1692. What kind of government does Mauritius have?
    Parliamentary republic

1693. Dutch colonization started in 1638 and ended in 1710, with a brief interruption during which period?
1658–1666
1694. Mauritius was a French colony during which period?
1710–1810
1695. Mauritius was captured on December 3rd, 1810 by which country?
United Kingdom
1696. When did Mauritius gain independence from United Kingdom?
March 12th, 1968
1697. When did Mauritius become a republic within the Commonwealth?
March 12th, 1992
1698. What is the area of Mauritius?
787 sq mi / 2,040 km^2
1699. How long is the coastline of Mauritius?
110 mi / 177 km
1700. What is the population of Mauritius?
1,284,264 (by 2009)
1701. What is the currency of Mauritius?
Mauritian Rupee
1702. What is the geographical feature of the terrain of Mauritius?
Small coastal plain rising to discontinuous mountains encircling central plateau
1703. What is the highest point of Mauritius?
Mont Piton (2,717 ft / 828 m)
1704. What is the lowest point of Mauritius?
Indian Ocean (0 ft / 0 m)
1705. Mauritius is part of which Archipelago?
Mascarene Islands
1706. Mascarene Islands is comprised of which islands?
Mauritius, Réunion (France), Rodrigues, Cargados Carajos shoals, plus the former islands of Saya de Malha Bank, Nazareth Bank and Soudan Bank
1707. What is special about Mascarene Islands?
They share a common geologic origin in the volcanism of the Réunion hotspot beneath the Mascarene Plateau and form a distinct ecoregion with a unique flora and fauna
1708. The island of Mauritius is renowned for having been the only known home of which flightless bird?
Dodo
1709. The island of Mauritius is of volcanic origin and is almost entirely surrounded by what?
Coral reefs
1710. Which island is a part of Mascarene Islands and a dependency lying 350 miles (560 km) northeast of Mauritius?
Rodrigues
1711. How large is Rodrigues?
42 sq mi / 109 km^2
1712. What is Cargados Carajos Shoals?

A dependency, lying 250 miles (402 km) north of Mauritius, with 16 small islands and islets on an extended reef in the Indian Ocean northeast of Mauritius

1713. How large is Agalega Islands, a dependency lying 700 miles (1,100 km) north of Mauritius?
9 sq mi / 24 km²
1714. What are two islands of Agalega Islands?
North Island (5.5 sq mi / 14.3 km²) and South Island (3.7 sq mi / 9.7 km²)
1715. What is Agalega Islands famous for?
Coconuts and Agalega Island Day Gecko
1716. Mauritius claims Tromelin Island as a dependency that is currently in which country's possession?
France
1717. Mauritius claims Chagos Archipelago as a dependency that is currently in which country's possession?
United Kingdom
1718. Mauritius contains what national parks?
Black River Gorges National Park, Pigeon Rock National Park, Ilot Vacoas National Park, Ile d'Ambre National Park, Ile aux Oiseaux National Park, Ilot Fous National Park, Ilot Fouquets National Park, Ile aux Flamants National Park, and Rocher aux Oiseaux National Park
1719. What are administrative divisions called in Mauritius?
Districts
1720. How many districts does Mauritius have?
9 (plus 3 dependencies)
1721. What is the climate of Mauritius?
Tropical, modified by southeast trade winds; warm, dry winter (May to November); hot, wet, humid summer (November to May)
1722. What are the natural resources of Mauritius?
Arable land, fish
1723. What are the natural hazards of Mauritius?
Cyclones (November to April); almost completely surrounded by reefs that may pose maritime hazards
1724. What are the religions of Mauritius?
Hindu 48%, Roman Catholic 23.6%, Muslim 16.6%, other Christian 8.6%, other 2.5%, unspecified 0.3%, none 0.4%
1725. What are the ethnic groups of Mauritius?
Indo-Mauritian 68%, Creole 27%, Sino-Mauritian 3%, Franco-Mauritian 2%

Mayotte (France)

1726. What is Mayotte?
Overseas collectivity of France
1727. Mayotte is part of which archipelago?
Comoro Islands
1728. Mayotte is located in which channel in Indian Ocean?
Mozambique Channel in the Indian Ocean

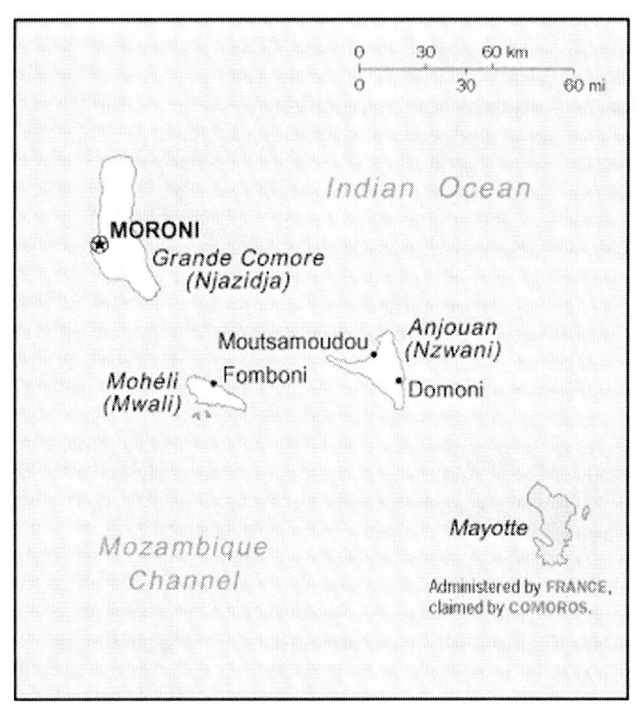

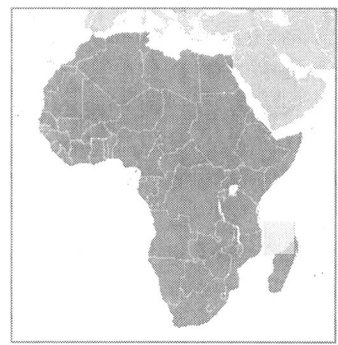

1729. Mayotte is between which two countries?
    Madagascar and Mozambique
1730. What is the motto of Mayotte?
    Liberty, Equality, Fraternity
1731. What is the national anthem of Mayotte?
    La Marseillaise
1732. What is the capital of Mayotte?
    Mamoudzou (the largest city)
1733. What is the official language of Mayotte?
    French
1734. Why does France still administer Mayotte as an overseas collectivity after the independence of Comoros even though it is part of the Comoros archipelago?
    Because Mayotte was the only island in the archipelago that voted against independence from France, and France has vetoed United Nations Security Council resolutions that would affirm Comorian sovereignty over the island
1735. A March 29th, 2009 referendum decided that Mayotte would become an overseas department of France in which year?
    2011
1736. What is the area of Mayotte?
    144 sq mi / 374 km^2
1737. How long is the coastline of Mayotte?
    115 mi / 185 km
1738. What is the population of Mayotte?
    223,765 (by 2009)
1739. What is the currency of Mayotte?

Euro
1740. What is the geographical feature of the terrain of Mayotte?
Generally undulating, with deep ravines and ancient volcanic peaks
1741. Mayotte is composed of a main island, a smaller island, and several islets around. What is name of the main island?
Grande-Terre (also called Mahoré)
1742. What is the name of the smaller island?
Petite-Terre (also called Pamanzi)
1743. What is the highest point of Mayotte?
Benara (2,165 ft / 660 m)
1744. What is the lowest point of Mayotte?
Indian Ocean (0 ft / 0 m)
1745. What are administrative divisions called in Mayotte?
Communes
1746. How many communes does Mayotte have?
17
1747. What is the climate of Mayotte?
Tropical; marine; hot, humid, rainy season during northeastern monsoon (November to May); dry season is cooler (May to November)
1748. What are the natural resources of Mayotte?
Negligible
1749. What are the natural hazards of Mayotte?
Cyclones during rainy season
1750. What are the religions of Mayotte?
Muslim 97%, Christian (mostly Roman Catholic) 3%

Morocco

1751. What is the official name of Morocco?
Kingdom of Morocco
1752. Which water body borders Morocco to the west?
Atlantic Ocean
1753. Which water body borders Morocco to the north?
Mediterranean Sea
1754. Which country borders Morocco to the east?
Algeria
1755. Which territory borders Morocco to the south?
Western Sahara
1756. How does Spain border Morocco?
A water border through the Strait of Gibraltar and land borders with three small Spanish enclaves, Ceuta, Melilla, and Peñón de Vélez de la Gomera
1757. What islands belong to Spain off the Atlantic coast of Morocco?
Canary Islands

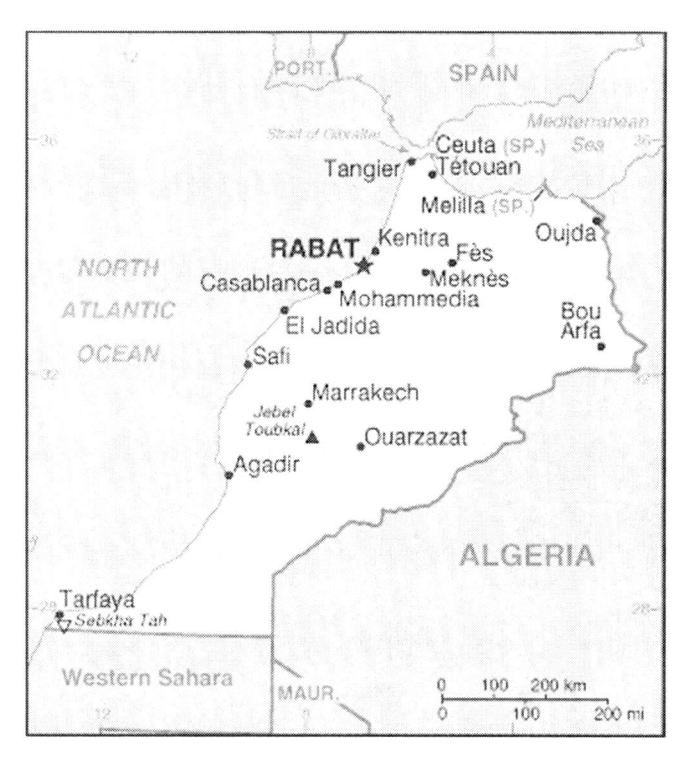

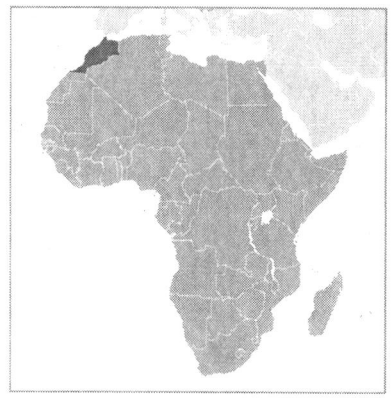

1758. What is the Strait of Gibraltar?
A narrow strait that connects the Atlantic Ocean to the Mediterranean Sea and separates Spain from Morocco
1759. What is Cape Spartel?
A promontory in Morocco about 305 mi / 1,000 ft above sea level at the entrance to the Strait of Gibraltar
1760. Is Cape Spartel the northernmost point of Africa?
No, it is Ras ben Sakka (Tunisia)
1761. What is the motto of Morocco?
God, Nation, King
1762. What is the national anthem of Morocco?
Hymne Chérifien
1763. What is the capital of Morocco?
Rabat
1764. What is the largest city of Morocco?
Casablanca
1765. What is the official language of Morocco?
Arabic
1766. What kind of government does Morocco have?
Constitutional monarchy
1767. Morocco was annexed to which empire in 40 CE?
Roman Empire
1768. Who conquered Morocco in the 7th century?
Arabs

1769. What was Barghawata Confederacy during 744–1058?
    A medieval Berber tribe confederation of the Atlantic coast of Morocco
1770. What was the Religion of Barghawata Confederacy?
    Islam
1771. What was the capital of Idrisid Dynasty, a state in western Maghreb (refer collectively to the countries of Morocco, Algeria, and Tunisia) during 780–974?
    Fez
1772. What was the capital of Almoravid Dynasty, Berber dynasty of Sahara during 1040–1147?
    Aghmat (1040-1062), Marrakech (1062-1147) and Córdoba (1147)
1773. What was Almohad Dynasty during 1121–1269?
    A Berber Dynasty that conquered all of northern Africa as far as Libya, together with Al-Andalus (Moorish Iberia of present-day southern Spain and Portugal)
1774. Almohad Empire was preceded by which countries?
    Almoravid Dynasty, Zirid Dynasty (present-day eastern Algeria and Tunisia), and Hammadid (present-day Algeria)
1775. Almohad Empire was succeeded by which countries?
    Taifa Emirate (Moorish Iberia of present-day southern Spain and Portugal), Hafsid Dynasty (present-day Tunisia, eastern Algeria and western Libya), Abdalwadid Dynasty (present-day Algeria), Emirate of Granada (present-day Spain), and Marinid Dynasty
1776. What was the capital of Marinid Dynasty during 1215–1465?
    Fez
1777. What was Kingdom of Fez (also called Wattassid sultanate) during 1472–1554?
    A state of North Africa, bordering the Mediterranean Sea to its north, Melilla (Spain) to the northeast, Abdalwadid dynasty (Present-day Algeria) to the east, Saadi Dynasty by Oum Er-Rbia River to the south, Atlantic ocean to the west, and Ceuta (Portugal) to the northwest
1778. What was the capital of Kingdom of Fez?
    Fez
1779. What was Saadi Dynasty during 1509–1659?
    A state in the south of the present-day Morocco
1780. What was the capital of Saadi Dynasty?
    Marrakech
1781. What is Alaouite Dynasty?
    The current Moroccan royal family starting from 1666
1782. What was the First Moroccan Crisis (also called Tangier Crisis)?
    The international crisis over the status of Morocco as Germany attempted to prevent France from establishing a protectorate over Morocco
1783. Which country supported Germany at the First Moroccan Crisis?
    Austria-Hungary
1784. Which courtiers supported France at the First Moroccan Crisis?
    United Kingdom, Russia, Italy, and Spain
1785. During which period did the First Moroccan Crisis happen?
    March 1905–May 1906
1786. What was Algeciras Conference?
    A conference held in Algeciras (Spain), from January 16th to April 7th, 1906 to find an issue to the First Moroccan Crisis

1787. What was the result of Algeciras Conference?
Moroccan Police Forces were founded under Spanish and French control
1788. What was the Second Moroccan Crisis (also called Agadir Crisis or Panther Sprung)?
The international crisis sparked by the deployment of the German gunboat Panther, to the Moroccan port of Agadir on July 1, 1911.
1789. Which country supported Germany at the Second Moroccan Crisis?
Austria-Hungary
1790. Which courtiers supported France at the Second Moroccan Crisis?
United Kingdom, Russia, Italy, and Spain
1791. When was Treaty of Fez signed?
March 30th, 1912?
1792. What was the result of Treaty of Fez?
Germany accepted France's position in Morocco in return for territory in the French Equatorial African colony of Middle Congo Germany
1793. What was the capital of French protectorate of Morocco?
Rabat
1794. What was the religion of French Protectorate of Morocco?
Catholicism, Islam
1795. What was the currency of French Protectorate of Morocco during 1912–1921?
Moroccan Rial
1796. What was the currency of French Protectorate of Morocco during 1921–1956?
Moroccan Franc
1797. When did Morocco gain independence from France?
March 2nd, 1956
1798. What was the result of Spanish-Moroccan War during October 22nd, 1859–April 26th, 1860?
Morocco recognized Spanish sovereignty over Ceuta and Melilla, ceded Sidi Ifni to Spain, and paid war reparations to Spain
1799. What was First Rif War (also called Melilla War or Magallo War)?
A conflict between Spain and Rif tribes of northern Morocco, and later the Sultan of Morocco
1800. When did First Rif War happen?
November 9th (de facto October 3rd), 1893–April 25th, 1894
1801. How was First Rif War resolved?
Melilla hinterlands ceded to Spain
1802. What was Second Rif War?
A conflict around Melilla between Spain and Rif tribes of northern Morocco
1803. When was Second Rif War?
1909
1804. What was the result of Second Rif War?
Spanish expanded their Melilla enclave
1805. When was Spanish Protectorate of Morocco established in northern Morocco?
February 27th, 1913
1806. What was the capital of Spanish Protectorate of Morocco?
Tétouan
1807. What was the religion of Spanish Protectorate of Morocco?

Catholicism, Islam
1808. What was the currency of Spanish Protectorate of Morocco?
Spanish Peseta
1809. What was Republic of the Rif?
A country created by Rif tribes revolted and declared their independence from Spanish occupation as well as from the Moroccan sultan
1810. What was the full name of Republic of the Rif?
The Confederal Republic of the Tribes of the Rif
1811. When was Republic of the Rif established?
September 18th, 1921
1812. What was the capital of Republic of the Rif?
Ajdir
1813. What was the religion of Republic of the Rif?
Muslim
1814. What was the currency of Republic of the Rif?
Rif Republic Riffan
1815. When was Republic of the Rif disestablished?
May 27th, 1926
1816. What caused Republic of the Rif to disestablish?
Third Rif War (also called the Second Moroccan War) in 1920
1817. When did Morocco gain independence from Spain?
April 7th, 1956
1818. What was Ifni War (also called Forgotten War)?
A series of armed incursions into Spanish West Africa by Moroccan insurgents
1819. During which period did Ifni War happen?
October 23rd, 1957–June 30th, 1958
1820. Which country was Spain's ally during the Ifni War?
France
1821. What was the result of Ifni War?
Morocco recognized Spanish rights to Ifni and the Spanish Sahara, and Cabo Juby ceded to Morocco
1822. When did Spain return Ifni to Morocco?
1969
1823. What was Perejil Island Crisis?
An armed conflict between Spain and Morocco over the small, uninhabited Perejil Island
1824. When did Perejil Island Crisis happen?
July 18th, 2002
1825. During Perejil Island Crisis, Spain was supported by all European Union countries except which country?
France
1826. During Perejil Island Crisis, Morocco was supported by all Arab League states except which country?
Algeria
1827. What was the result of Perejil Island Crisis?
Perejil Island remains unoccupied but claimed by either side

1828. What was Sand War?
Along the Algerian-Moroccan border, Moroccan attempted to claim some former French colony
1829. When did Sand War happen?
October 1963
1830. What was the result of Sand War?
The closing of part of the Algerian-Moroccan border
1831. What is the area of Morocco?
172,414 sq mi / 446,550 km^2
1832. How long is the coastline of Morocco?
1140 mi / 1,835 km
1833. What is the population of Morocco?
31,285,174 (by 2009)
1834. What is the currency of Morocco?
Moroccan Dirham
1835. What is the geographical feature of the terrain of Morocco?
Northern coast and interior are mountainous with large areas of bordering plateaus, intermontane valleys, and rich coastal plains
1836. What is the highest point of Morocco?
Jebel Toubkal (13,665 ft / 4,165 m)
1837. Jebel Toubkal is in which national park in Morocco?
Toubkal National Park (147 sq mi / 380 km^2)
1838. Besides Toubkal National Park, what are other national parks in Morroco?
Al Hoceima National Park, Ayachi National Park, Bou Arfa National Park, Dakhla National Park, Golf of Khnifiss National Park, Ifrane National Park, Iles of Essaouira National Park, Lake Sidi Boughaba National Park, Merdja Zerka National Park, Souss-Massa National Park, Talassemtane National Park, Tazekka National Park, and Tazerkourt National Park
1839. Jebel Toubkal is a peak in which mountain range?
Atlas Mountains
1840. Is Rif mountains part of the Atlas mountain range?
No, Atlas Mountains is located mainly in the center-southern Morocco, but Rif Mountains is located in the northern Morocco
1841. What is the lowest point of Morocco?
Sebkha Tah (-180 ft / -55 m)
1842. What is the longest river in Morocco?
Oum Er-Rbia River (345 mi / 555 km)
1843. What is Sebou River's rank by water flow in Morocco?
2nd (25 cu mi/s or 105 m^3/s)
1844. What is the second longest river in Morocco?
Sebou River (285 mi / 555 km)
1845. What is Sebou River's rank by water flow in Morocco?
1st (33 cu mi/s or 137 m^3/s)
1846. What are administrative divisions called in Morocco?
Regions
1847. How many regions does Morocco have?

1848. What is the climate of Morocco?
Mediterranean, becoming more extreme in the interior
1849. What are the natural resources of Morocco?
Phosphates, iron ore, manganese, lead, zinc, fish, salt
1850. What are the natural hazards of Morocco?
Northern mountains geologically unstable and subject to earthquakes; periodic droughts
1851. What are the religions of Morocco?
Muslim 98.7%, Christian 1.1%, Jewish 0.2%
1852. What are the ethnic groups of Morocco?
Arab-Berber 99.1%, other 0.7%, Jewish 0.2%

Mozambique

1853. What is the official name of Mozambique?
Republic of Mozambique
1854. Which countries border Mozambique to the northwest?
Malawi and Zambia
1855. Which countries border Mozambique to the southwest?
Swaziland and South Africa
1856. Which country borders Mozambique to the north?
Tanzania
1857. Which country borders Mozambique to the west?
Zimbabwe
1858. Which water body borders Mozambique to the east?
Indian Ocean
1859. What is the national anthem of Mozambique?
Lovely Homeland
1860. What is the capital of Mozambique?
Maputo (the largest city)
1861. What is the official language of Mozambique?
Portuguese
1862. What kind of government does Mozambique have?
Republic
1863. Why is Mozambique's membership in the Commonwealth unique?
They were the first country to join without any former ties with United Kingdom
1864. In which year did Mozambique become the colony of Portugal?
1498
1865. What were common names of Mozambique as Portuguese colony across different periods of time?
Portuguese East Africa, Portuguese Mozambique or Overseas Province of Mozambique
1866. What was the currency of Mozambique during Portuguese rule?
Mozambican Escudo
1867. When did Mozambican War of Independence start?
September 25th, 1964

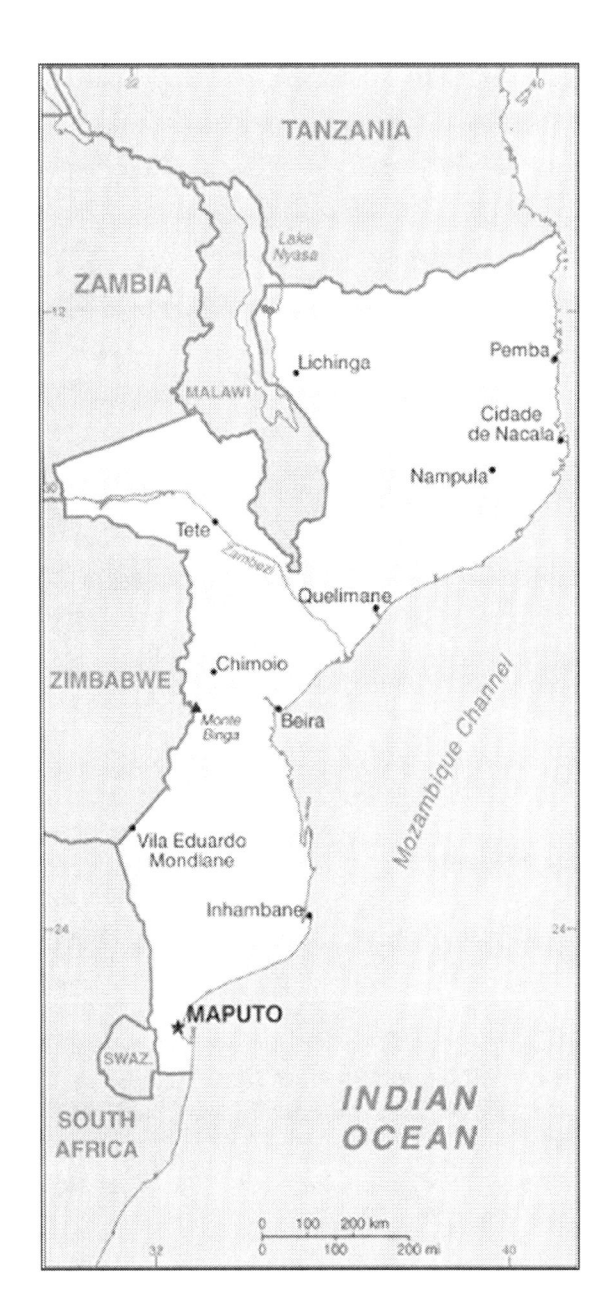

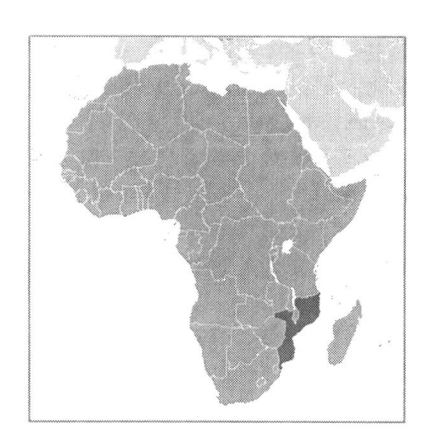

1868. When did Mozambican War of Independence cease fire?
	September 8th, 1974
1869. When did Mozambique gain independence from Portugal?
	June 25th, 1975
1870. When did Mozambican Civil War start?
	1977
1871. When did Mozambican Civil War end?
	1992
1872. What were results of Mozambican Civil War?
	Rome General Peace Accords and multiparty elections

1873. When was Rome General Peace Accords signed between the Mozambican Civil War parties
October 4th, 1992
1874. What is the area of Mozambique?
308,642 sq mi / 799,380 km²
1875. How long is the coastline of Mozambique?
1,535 mi / 2,470 km
1876. What is the population of Mozambique?
21,669,278 (by 2009)
1877. What is the currency of Mozambique?
Mozambican Metical
1878. What is the geographical feature of the terrain of Mozambique?
Mostly coastal lowlands, uplands in center, high plateaus in northwest, mountains in west
1879. What is the highest point of Mozambique?
Monte Binga (7,992 ft / 2,436 m)
1880. Monte Binga is the highest peak of Chimanimani mountains in Mozambique, just east of the border with which country?
Zimbabwe
1881. Chimanimani Transfrontier Park is along the border of Mozambique and which country?
Zimbabwe
1882. What is the lowest point of Mozambique?
Indian Ocean (0 ft / 0 m)
1883. What is Maputo Bay (also called Delagoa Bay)?
An inlet of the Indian Ocean on the coast of Mozambique
1884. What is Inhaca Island?
A subtropical island of Mozambique off the East African coast, which separates Maputo Bay from the Indian Ocean
1885. How large is Inhaca Island?
20 sq mi / 52 km²
1886. Bazaruto Archipelago is a group of six islands in Mozambique. What are these islands?
Bazaruto, Benguerra, Magaruque, Banque, Santa Carolina (also called Paradise Island) and Shell
1887. Bazaruto Archipelago National Park (618 sq mi / 1,600 km²) was created in 1971. What are tourist attractions?
Sandy beaches, coral reefs, surfing and fishing
1888. Which river divides Mozambique into two topographical regions in the north and south
Zambezi River
1889. Which lakes are located between Malawi and Mozambique?
Lago Niassa (also called Lake Malawi, Lake Nyasa, or Lake Nyassa) and Lake Chiuta
1890. Which lake is located in Mozambique?
Lake Shirwa
1891. How large is Banhine National Park, located in southern Mozambique, which lies between the Limpopo River and the Changane River?
2,703 sq mi / 7,000 km²
1892. Gorongosa National Park (1,456 sq mi / 3,770 km²), located in central Mozambique, was featured by which movie on December 14th, 2009 in Maputo?

Gorongosa National Park: Africa's Lost Eden

1893. Which national park in Mozambique forms Great Limpopo Transfrontier Park, along with Gonarezhou National Park in Zimbabwe and Kruger National Park in South Africa?
Limpopo National Park (3,861 sq mi / 10,000 km²)

1894. How large is Quirimbas National Park, located in northern Mozambique, which comprises the 11 southernmost islands of the Quirimbas Archipelago?
2,896 sq mi / 7,500 km²

1895. How large is Zinave National Park, located in Inhambane Province of Mozambique, which has a diverse landscape of miombo forests, bushes, riverine vegetation, and lagoons?
2,317 sq mi / 6,000 km²

1896. What are administrative divisions called in Mozambique?
Provinces

1897. How many provinces does Mozambique have?
10 (plus 1 city)

1898. What is the climate of Mozambique?
Tropical to subtropical

1899. What are the natural resources of Mozambique?
Coal, titanium, natural gas, hydropower, tantalum, graphite

1900. What are the natural hazards of Mozambique?
Severe droughts; devastating cyclones and floods in central and southern provinces

1901. What are the religions of Mozambique?
Catholic 23.8%, Muslim 17.8%, Zionist Christian 17.5%, other 17.8%, none 23.1% (1997 census)

1902. What are the ethnic groups of Mozambique?
African 99.66% (Makhuwa, Tsonga, Lomwe, Sena, and others), Europeans 0.06%, Euro-Africans 0.2%, Indians 0.08%

Namibia

1903. What is the official name of Namibia?
Republic of Namibia

1904. Which country borders Namibia to the northeast?
Zambia

1905. Which country borders Namibia to the north?
Angola

1906. Which countries border Namibia to the east?
Botswana and Zimbabwe

1907. Which country borders Namibia to the south and east?
South Africa

1908. Which water body borders Namibia to the west?
Atlantic Ocean

1909. What is the motto of Namibia?
Unity, Liberty, Justice

1910. What is the national anthem of Namibia?
Namibia, Land of the Brave

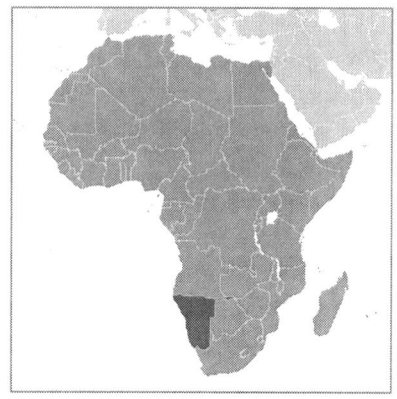

1911. What is the capital of Namibia?
Windhoek (the largest city)
1912. What is the official language of Namibia?
English
1913. What kind of government does Namibia have?
Republic
1914. When did German South West Africa exist as a colony of Germany?
1884–1915
1915. What was the currency of German South West Africa?
German South West African Mark
1916. Which country occupied South-West Africa during World War I and administered it as a League of Nations mandate territory?
South Africa
1917. When did South-West Africa exist under the rule of South Africa?
1915–1990
1918. What was the currency of South-West Africa during 1920–1961?
South West African Pound
1919. What was the currency of South-West Africa during 1961–1993?
South African Rand
1920. During which period was the Namibian War of Independence fought between nationalists and South Africa?
1966–1988
1921. When did Namibia gain independence from South Africa?
March 21st, 1990

1922. Which city remained under South African control until 1994
Walvis Bay
1923. What is the area of Namibia?
318,696 sq mi / 825,418 km^2
1924. How long is the coastline of Namibia?
977 mi / 1,572 km
1925. The Namibian government has proposed building which dam, a controversial hydroelectric dam on Kunene River (also called Cunene River)?
Epupa Dam
1926. What is the population of Namibia?
2,108,665 (by 2009)
1927. Is Namibia the least densely populated country in the world?
No, the second least densely populated country, after Mongolia
1928. What is the currency of Namibia?
Namibian Dollar
1929. What is the highest point of Namibia?
Konigstein (8,550 ft / 2,606 m)
1930. What is the lowest point of Namibia?
Atlantic Ocean (0 ft / 0 m)
1931. What is the geographical feature of the terrain of Namibia?
Mostly high plateau; Namib Desert along coast; Kalahari Desert in east
1932. What is world's oldest desert?
The Namib Desert (at least 55 million years old)
1933. How large is the Namib Desert?
31,200 sq mi / 80,900 km^2
1934. The Namib Desert covers which countries?
Namibia and Angola
1935. Which national park is an ecological preserve in the Namib Desert?
Namib-Naukluft National Park
1936. The dunes in the Namib Desert are the tallest in the world. How tall are they?
Almost 1,000 ft (more than 300 m) above the desert floor
1937. What is the narrow protrusion of Namibia eastwards?
Caprivi (also called Caprivi Strip, Okavango Strip, and Itenge)
1938. The Kalahari Desert covers Namibia and what other countries?
Botswana and South Africa
1939. What is Namibia's rank in the world for diamond production?
1st
1940. Where in Namibia is the richest source of diamonds on earth?
Namibia's coastal deserts
1941. What is Skeleton Coast?
The northern part of Namibia coast
1942. What is Diamond Coast?
The southern part of Namibia coast

1943. Ai-Ais/Richtersveld Transfrontier Park is a peace park straddling the border between South Africa and Namibia. Which national park in Namibia is part of Ai-Ais/Richtersveld Transfrontier Park?
Ai–Ais Hot Springs Game Park
1944. How large was Etosha National Park in 1907, located in northwestern Namibia, which was the largest game reserve in the world at that time?
38,500 sq mi / 100,000 km²
1945. How large is Etosha National Park, which excludes most of Kaokoland and the Etosha pan, part of the old Etosha National Park?
8,598 sq mi / 22,270 km²
1946. How large is Mamili National Park, centered on the Nkasa Island and Lupala Island of Namibia, where 80% of the area becomes flooded during the wet season?
124 sq mi / 320 km²
1947. How large is Skeleton Coast National Park, located in northwestern Namibia, which is dotted with shipwrecks?
7,722 sq mi / 20,000 km²
1948. How large is Waterberg National Park, located in central Namibia on the Waterberg Plateau, which has over 850 million year-old rock stratum and 200 million year-old dinosaur tracks?
156 sq mi / 405 km²
1949. What are administrative divisions called in Namibia?
Regions
1950. How many regions does Namibia have?
13
1951. What is the climate of Namibia?
Desert; hot, dry; rainfall sparse and erratic
1952. What are the natural resources of Namibia?
Diamonds, copper, uranium, gold, silver, lead, tin, lithium, cadmium, tungsten, zinc, salt, hydropower, fish (possible oil, coal, and iron ore deposits)
1953. What are the natural hazards of Namibia?
Prolonged periods of drought
1954. What are the religions of Namibia?
Christian 80% to 90% (Lutheran 50% at least), indigenous beliefs 10% to 20%
1955. What are the ethnic groups of Namibia?
Black 87.5%, white 6%, mixed 6.5%

Niger

1956. What is the official name of Niger?
Republic of Niger
1957. Niger was named after what?
Niger River
1958. Which country borders Niger to the east?
Chad

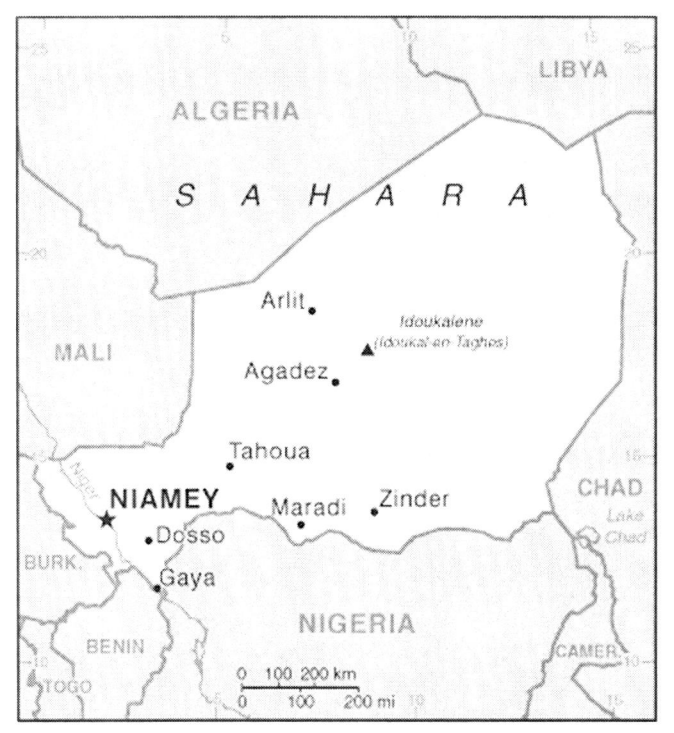

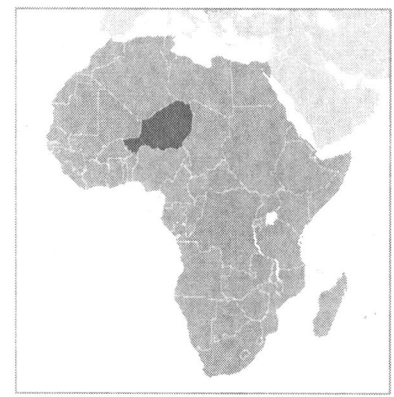

1959. Which countries border Niger to the southwest?
   Burkina Faso and Benin
1960. What country borders Niger to the south?
   Nigeria
1961. Which countries border Niger to the north?
   Algeria and Libya
1962. Which country borders Niger to the west?
   Mali
1963. What is the motto of Niger?
   Fraternity, Work, Progress
1964. What is the national anthem of Niger?
   La Nigérienne
1965. What is the capital of Niger?
   Niamey (the largest city)
1966. What is the official language of Niger?
   French
1967. What kind of government does Niger have?
   Presidential republic
1968. Was Niger part of Songhai Empire, Mali Empire, Dendi Kingdom, Wolof Empire, Gao Empire, and Kanem Empire?
   No, not Wolof Empire
1969. When did Dendi Kingdom exist?
   1591–1901
1970. What was the capital of Dendi Kingdom?

Lulama
1971. What was the language used in Dendi Kingdom?
Songhai
1972. What was the religion of Dendi Kingdom?
Islam
1973. Who conquered Dendi Kingdom in 1901?
France
1974. When did Niger become a French colony?
1922
1975. When did Niger gain independence from France?
August 3rd, 1960
1976. What is the area of Niger?
489,678 sq mi / 1,267,000 km^2
1977. How long is the coastline of Niger?
0 mi / 0 km
1978. What is the population of Niger?
15,306,252 (by 2009)
1979. What is the currency of Niger?
West African CFA Franc
1980. What is the geographical feature of the terrain of Niger?
Predominately desert plains and sand dunes; flat to rolling plains in south; hills in north
1981. What is the highest point of Niger?
Mont Bagzane (6,634 ft /2,022 m)
1982. Mont Bagzane is located in which mountain range?
Aïr Mountains
1983. Which was believed to be the highest peak in Niger until 2001?
Mont Gréboun (6,378 ft / 1,944 m)
1984. What is the lowest point of Niger?
Niger River (656 ft / 200 m)
1985. Ténéré is a desert region in the south central Sahara, comprising a vast plain of sand stretching from northeastern Niger into western Chad. How large is Ténéré Desert?
Over 154,440 sq mi (400,000 km^2)
1986. Which mountain is the west boundary of Ténéré Desert?
Aïr Mountains
1987. Which is the northeast boundary of Ténéré Desert?
Djado Plateau
1988. Niger and which two countries operate W National Park?
Benin and Burkina Faso
1989. What are administrative divisions called in Niger?
Regions
1990. How many regions does Niger have?
8 (plus 1 capital district)
1991. What is the climate of Niger?
Desert; mostly hot, dry, dusty; tropical in extreme south
1992. What are the natural resources of Niger?

Uranium, coal, iron ore, tin, phosphates, gold, molybdenum, gypsum, salt, petroleum
1993. What are the natural hazards of Niger?
Recurring droughts
1994. What are the religions of Niger?
Muslim 80%, other (includes indigenous beliefs and Christian) 20%
1995. What are the ethnic groups of Niger?
Haoussa 55.4%, Djerma Sonrai 21%, Tuareg 9.3%, Peuhl 8.5%, Kanouri Manga 4.7%, other 1.2%

Nigeria

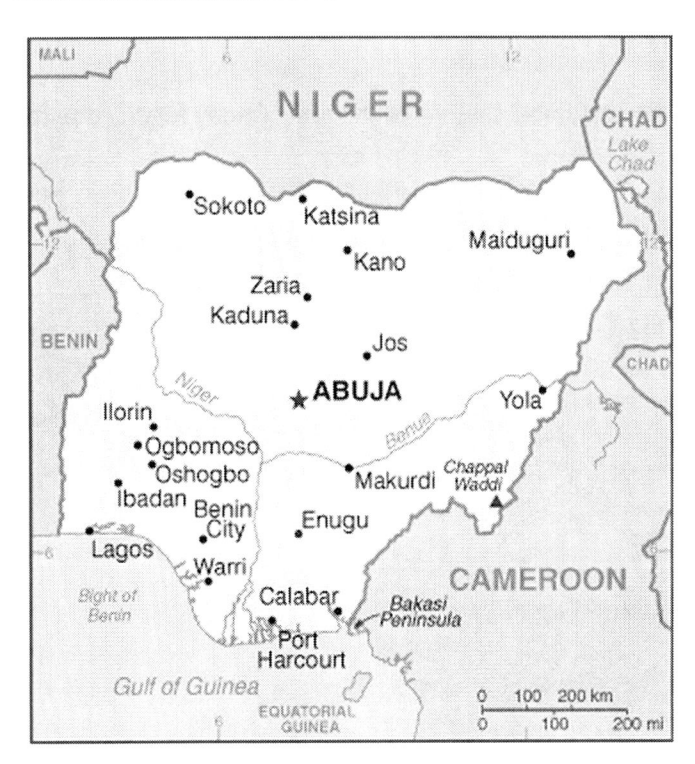

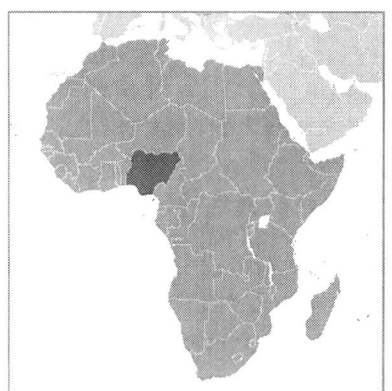

1996. What is the official name of Nigeria?
Federal Republic of Nigeria
1997. Which country borders Nigeria to the west?
Benin
1998. What country borders Nigeria to the northeast?
Chad
1999. Which country borders Nigeria to the north?
Niger
2000. Which country borders Nigeria to the east?
Cameroon
2001. Which water body borders Nigeria to the south?
Gulf of Guinea (in Atlantic Ocean)
2002. What is the motto of Nigeria?

Unity and Faith, Peace and Progress
2003. What is the national anthem of Nigeria?
Arise, O Compatriots
2004. What is the capital of Nigeria?
Abuja
2005. What is the largest city of Nigeria?
Lagos
2006. What is the official language of Nigeria?
English
2007. What kind of government does Nigeria have?
Presidential Federal Republic
2008. People lived in Nigeria as early as when?
9000 BCE
2009. What are early independent Kingdoms and states that make the present-day Nigeria due to British colonialism?
Calabar Empire, Ibibio Empire, Igbo Empire, Ijaw Empire, Kanem Empire, Bornu Empire, Nri Kingdom, Oyo Empire, Benin Empire, and Yoruba Kingdom
2010. When did Bornu Empire exist?
1387–1893
2011. What was the capital of Bornu Empire?
Ngazargamu
2012. What was the language used in Bornu Empire?
Kanuri
2013. What was the religion of Bornu Empire?
Islam
2014. How large was Bornu Empire?
193,051 sq mi / 500,000 km^2
2015. When did Oyo Empire exist in southwestern Nigeria?
1400–1905
2016. What was the capital of Oyo Empire?
Oyo-Ile
2017. What was the language used in Oyo Empire?
Yoruba
2018. What was the religion of Oyo Empire?
Yoruba religion
2019. How large was Oyo Empire?
57,915 sq mi / 150,000 km^2
2020. When did Nri Kingdom exist?
1043–1911
2021. When did Benin Empire (also called Edo Empire) exist?
1440–1897
2022. What was the capital of Benin Empire?
Ile-Ibinu
2023. What language was used in Benin Empire?
Edo

2024. Which country annexed Benin Empire in 1897?
United Kingdom
2025. Colonial Nigeria under United Kingdom started from when?
1800
2026. What was Oil Rivers Protectorate?
A British protectorate in the Oil Rivers area during 1885–1893
2027. Why was the Niger Delta region called the Oil Rivers?
Because it was a major producer of palm oil
2028. Oil Rivers Protectorate was expanded and became what?
Niger Coast Protectorate
2029. Niger Coast Protectorate merged with the chartered territories of the Royal Niger Company to form what?
Southern Nigeria
2030. What was Southern Nigeria?
A British protectorate in the coastal areas of present-day Nigeria
2031. When did Southern Nigeria exist?
January 1st, 1900–1 January 1st, 1914
2032. Areas around Lagos was added Southern Nigeria in 1906, and the territory was officially renamed to what?
Colony and Protectorate of Southern Nigeria
2033. When did Northern Nigeria exist?
January 1st, 1900–1 January 1st, 1914
2034. Why did Southern Nigeria join with Northern Nigeria to form the single colony of Nigeria?
To use the budget surpluses in Southern Nigeria to offset deficit in Northern Nigeria
2035. When did Nigeria gain independence from United Kingdom?
October 1st, 1960
2036. During which period did Nigerian Civil War (also called Nigerian-Biafran War) happen?
July 6th, 1967–January 15th, 1970
2037. What kind of conflict caused Nigerian Civil War?
The attempted secession of the southeastern provinces of Nigeria as the self-proclaimed Republic of Biafra
2038. Was Republic of Biafra a recognized country?
No, it was only recognized by Gabon, Haiti, Côte d'Ivoire, Tanzania and Zambia
2039. What was the motto of Biafra?
Peace, Unity, Freedom
2040. What was the national anthem of Biafra?
Land of the Rising Sun
2041. What was the capital Biafra?
Enugu
2042. What was the currency of Biafra?
Biafran Pound
2043. What was the result of Nigerian Civil War?
Biafra ended on January 15th, 1970
2044. What is the area of Nigeria?
356,667 sq mi / 923,768 km^2

2045. How long is the coastline of Nigeria?
530 mi / 853 km
2046. What is the population of Nigeria?
149,229,090 (by 2009)
2047. What is Nigeria's rank by population in Africa?
1st
2048. What is the currency of Nigeria?
Nigerian Naira
2049. What is the geographical feature of the terrain of Nigeria?
Southern lowlands merge into central hills and plateaus; mountains in southeast, plains in north
2050. What is the highest point of Nigeria?
Chappal Waddi (7,936 ft / 2,419 m)
2051. What is the lowest point of Nigeria?
Atlantic Ocean (0 ft / 0 m)
2052. What is the most important landmark of Kainji Lake National Park (2,062 sq mi / 5,340 km^2)?
Kainji Lake
2053. Kainji Lake, located in western Nigeria, is a reservoir of which river?
Niger River
2054. Kainji Lake was formed by which dam, constructed from 1964 to 1968?
Kainji Dam
2055. Which national park contains the ruins of the original Oyo city?
Old Oyo National Park
2056. What is the biggest national park in northeastern Nigeria?
Yankari National Park (870 sq mi / 2,244 km²)
2057. What is Adamawa Plateau?
A plateau region in west-central Africa stretching from southeastern Nigeria through north-central Cameroon to the Central African Republic
2058. What is Jos Plateau?
A plateau located near the center of Nigeria
2059. Niger River and Benue River converge and empty into which delta?
Niger Delta
2060. Niger River valley and Benue River valley merge into each other and form what shape?
Y
2061. Which city in Cross River State is believed to contain the world's largest diversity of butterflies?
Calabar
2062. What is the longest river in Niger?
Niger River
2063. What are administrative divisions called in Nigeria?
States
2064. How many states does Nigeria have?
36 (plus 1 territory)
2065. What is the climate of Nigeria?

Varies; equatorial in south, tropical in center, arid in north
2066. What are the natural resources of Nigeria?
Natural gas, petroleum, tin, iron ore, coal, limestone, niobium, lead, zinc, arable land
2067. What are the natural hazards of Nigeria?
Periodic droughts; flooding
2068. What are the religions of Nigeria?
Muslim 50%, Christian 40%, indigenous beliefs 10%
2069. What are the ethnic groups of Nigeria?
Hausa and Fulani 29%, Yoruba 21%, Igbo (Ibo) 18%, Ijaw 10%, Kanuri 4%, Ibibio 3.5%, Tiv 2.5%

Réunion (France)

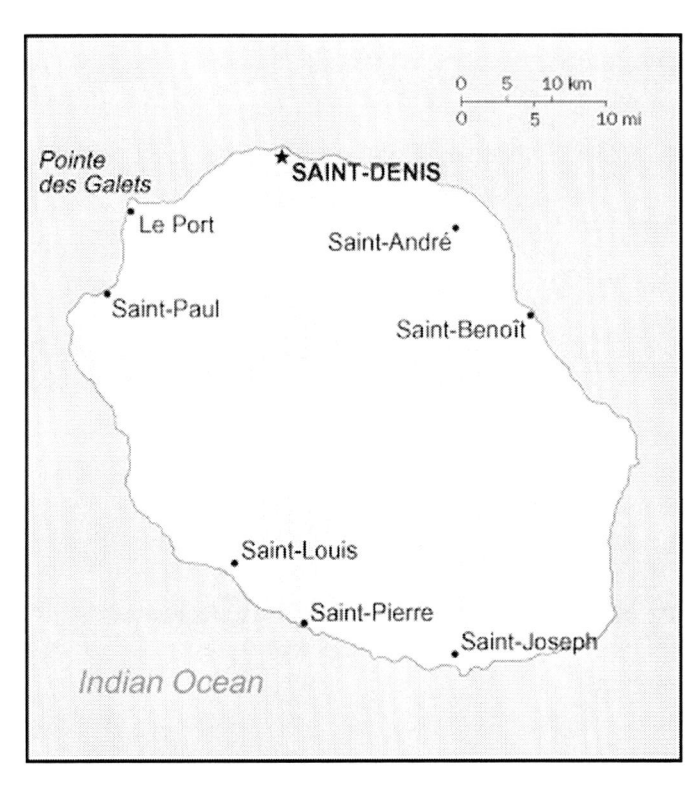

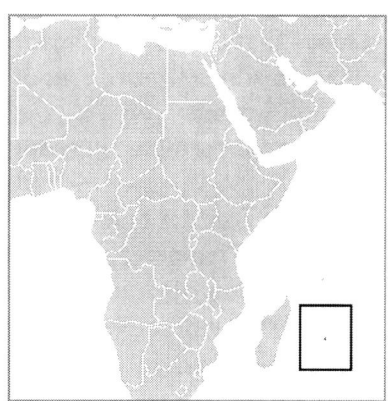

2070. What is Réunion?
Overseas department of France
2071. Réunion is an island in which water body?
Indian Ocean
2072. Réunion is located about 120 miles (200 km) southwest of which country?
Mauritius
2073. Réunion belongs to which archipelago?
Mascarene Islands
2074. What is the capital of Réunion?
Saint-Denis (the largest city)

2075. What is the official language of Réunion?
French
2076. What is the area of Réunion?
970 sq mi / 2,512 km²
2077. How long is the coastline of Réunion?
129 mi / 207 km
2078. What is the population of Réunion?
827,000 (by 2009)
2079. What is the currency of Réunion?
Euro
2080. What is the geographical feature of the terrain of Réunion?
Mostly rugged and mountainous; fertile lowlands along coast
2081. What is the highest point of Réunion?
Piton des Neiges (10,069 ft / 3,069 m)
2082. How high is Piton de la Fournaise?
8,630 ft / 2,631 m
2083. What type of mountains is Piton des Neiges or Piton de la Fournaise?
Shield volcano
2084. When was the recent eruption of Piton de la Fournaise?
January 12th, 2010
2085. What is the lowest point of Réunion?
Indian Ocean (0 ft / 0 m)
2086. What was the impact of the Suez Canal opening in November 1869?
It made the island lose its importance as a stopover on the East Indies trade route
2087. What are administrative divisions called in Réunion?
Arrondissements
2088. How many arrondissements does Réunion have?
4
2089. What is the climate of Réunion?
Tropical, but temperature moderates with elevation; cool and dry (May to November), hot and rainy (November to April)
2090. What are the natural resources of Réunion?
Fish, arable land, hydropower
2091. What are the natural hazards of Réunion?
Periodic, devastating cyclones (December to April); Piton de la Fournaise on the southeastern coast is an active volcano
2092. What are the religions of Réunion?
Roman Catholic 86%, Hindu, Muslim, Buddhist
2093. What are the ethnic groups of Réunion?
French, African, Malagasy, Chinese, Pakistani, Indian

Plazas de soberanía (Spain)

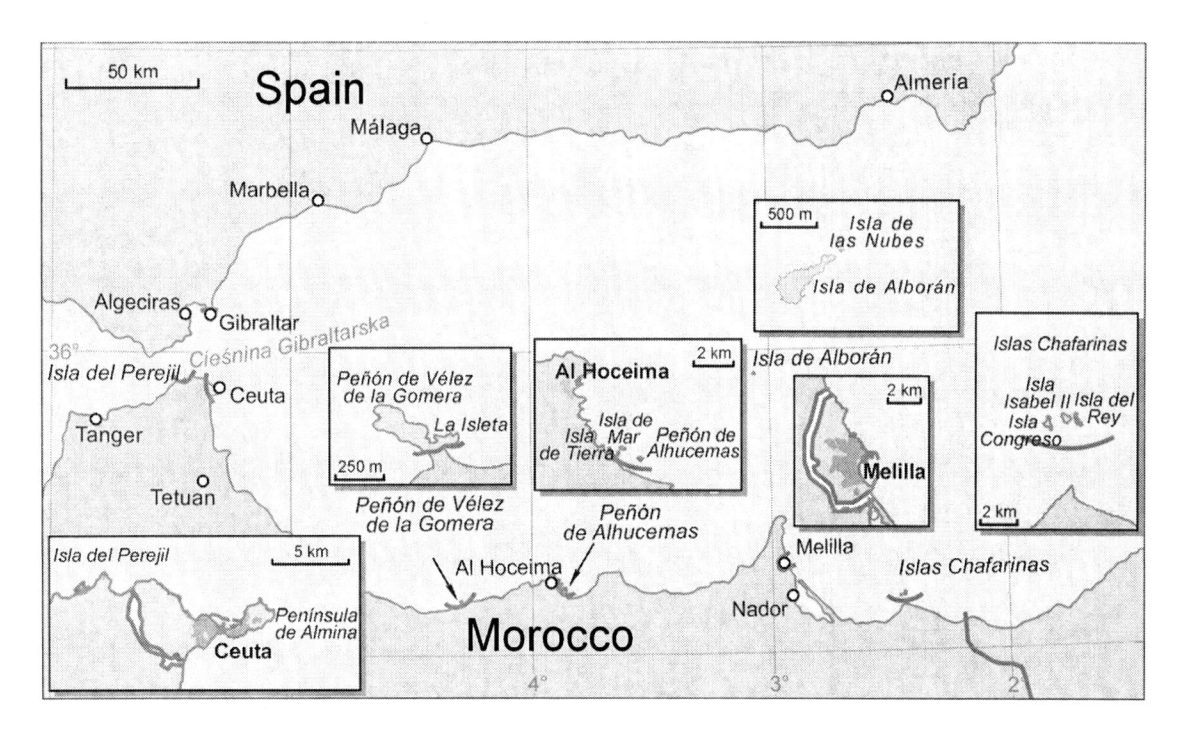

2094. What is Plazas de soberanía ("places of sovereignty")?
It refers to Spanish North Africa, the current Spanish territories in continental North Africa, bordering Morocco
2095. What are included in Plazas de soberanía?
2 autonomous cities of Ceuta and Melilla (large places of sovereignty), and the Islas Chafarinas, the Peñón de Alhucemas and the Peñón de Vélez de la Gomera (lesser sovereignty places)
2096. Which small uninhabited islet close to Ceuta has been lately defined as an extra plaza de soberanía?
Isla Perejil
2097. Which small island is in the western Mediterranean, about 31 mi / 50 km north of the Moroccan coast and 56 mi / 90 km south of Spain?
Isla de Alborán
2098. Plazas de soberanía borders which water body?
The Mediterranean Sea
2099. Plazas de soberanía is also claimed by which other country?
Morocco
2100. What is the official language of Plazas de soberanía?
Spanish
2101. What is the currency of Plazas de soberanía?
Euro
2102. What separates Ceuta from Spain mainland?
The Strait of Gibraltar
2103. What is the area of Ceuta?
7.5 sq mi / 19.5 km^2

2104. What is the population of Ceuta?
77,389 (by 2008)
2105. What is the area of Melilla?
4.7 sq mi / 12.3 km^2
2106. What is the population of Melilla?
71,448 (by 2008)

Rwanda

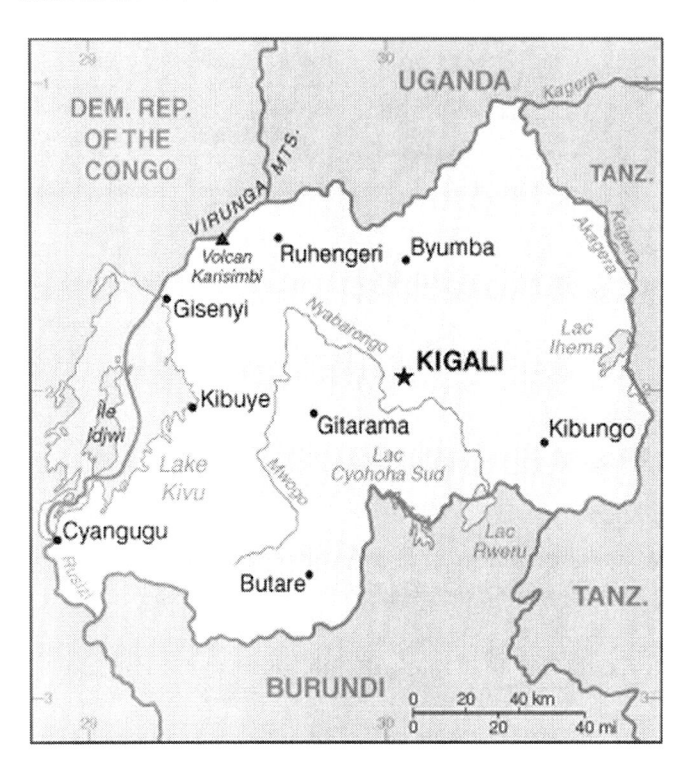

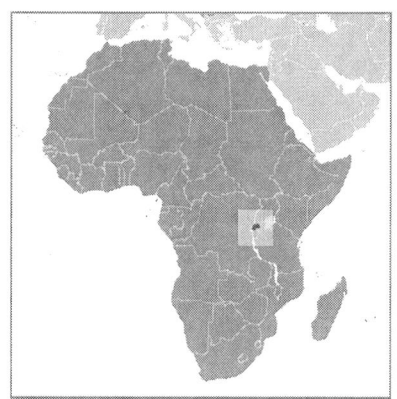

2107. What is the official name of Rwanda?
Republic of Rwanda
2108. Which country borders Rwanda to the north?
Uganda
2109. Which country borders Rwanda to the east?
Tanzania
2110. Which country borders Rwanda to the south?
Burundi
2111. Rwanda is separated from Democratic Republic of the Congo the Ruzizi River valley to the west and by which lake?
Lake Kivu
2112. What is the motto of Rwanda?
Unity, Work, Patriotism
2113. What is the national anthem of Rwanda?
Beautiful Rwanda

2114. What is the capital of Rwanda?
    Kigali (the largest city)
2115. What are the official languages of Rwanda?
    Kinyarwanda, French, English
2116. What kind of government does Rwanda have?
    Republic
2117. Rwanda was part of which colony during February 27th, 1885–June 28th, 1919?
    German East Africa
2118. What was Ruanda-Urundi?
    A Belgian suzerainty from 1916 to 1924, a League of Nations Class B Mandate from 1924 to 1945 and then a UN trust territory until 1962
2119. What was the capital of Ruanda-Urundi?
    Bujumbura (Burundi)
2120. What was the official language of Ruanda-Urundi?
    French
2121. When did Rwanda gain independence from Belgium?
    July 1st, 1962
2122. What was the Rwandan Civil War?
    A conflict between the government of President Juvénal Habyarimana and the rebel Rwandan Patriotic Front
2123. When did Rwandan Civil War begin?
    October 2nd, 1990
2124. Rwandan Civil War ended on August 4th, 1993 with the signing of which agreements?
    Arusha Accords (create a power-sharing government)
2125. Which event on the evening of April 6, 1994 was the catalyst for Rwandan Genocide?
    The assassination of Juvénal Habyarimana
2126. What was Rwandan Genocide in 1994?
    The mass killing of hundreds of thousands of Rwanda's Tutsis and Hutu political moderates by the Hutu dominated government
2127. What was the estimated death roll of Rwandan Genocide?
    800,000 or 20% of the total population
2128. When did Rwanda join the Commonwealth?
    November 29th, 2009
2129. What was the significance of Rwanda's membership of the Commonwealth?
    It was the second nation to join the Commonwealth without former ties with United Kingdom
2130. What is the area of Rwanda?
    10,169 sq mi / 26,338 km^2
2131. How long is the coastline of Rwanda?
    0 mi / 0 km
2132. What is the population of Rwanda?
    10,264,947 (by 2010)
2133. What is the currency of Rwanda?
    Rwandan Franc
2134. What is the geographical feature of the terrain of Rwanda?

Mostly grassy uplands and hills; relief is mountainous with altitude declining from west to east

2135. Because of its hilly geography, what is Rwanda known as?
The Land of a Thousand Hills

2136. What is the highest point of Rwanda?
Volcan Karisimbi (14,859 ft / 4,519 m)

2137. Volcan Karisimbi is the highest in which mountain range that includes a chain of volcanoes in East Africa?
Virunga Mountains

2138. Virunga Mountains is located in Rwanda and which two other countries?
Democratic Republic of the Congo and Uganda

2139. What is the lowest point of Rwanda?
Rusizi River (3,117 ft / 950 m)

2140. How large is Akagera National Park in north eastern Rwanda, which lies against the Tanzanian border?
965 sq mi / 2,500 km²

2141. Which lakes are included in Akagera National Park?
Lake Shakani and Lake Ihema

2142. How large is Nyungwe Forest National Park in southwestern Rwanda, which is located south of Lake Kivu on the border with Burundi?
375 sq mi / 970 km²

2143. How large is Volcanoes National Park in northwestern Rwanda?
50 sq mi / 130 km²

2144. Which five of the eight volcanoes of the Virunga Mountains are located in Volcanoes National Park?
Volcan Karisimbi, Volcan Bisoke (12,175 ft / 3,711 m), Volcan Muhabura (13,540 ft / 4,127 m), Volcan Gahinga (11,398 feet / 3,474 m), and Volcan Sabyinyo (11,959 ft / 3,645 m)

2145. What are administrative divisions called in Rwanda?
Provinces

2146. How many provinces does Rwanda have?
4

2147. What is the climate of Rwanda?
Temperate; two rainy seasons (February to April, November to January); mild in mountains with frost and snow possible

2148. What are the natural resources of Rwanda?
Gold, cassiterite (tin ore), wolframite (tungsten ore), methane, hydropower, arable land

2149. What are the natural hazards of Rwanda?
Periodic droughts; the volcanic Virunga Mountains

2150. What are the religions of Rwanda?
Roman Catholic 56.5%, Protestant 26%, Adventist 11.1%, Muslim 4.6%, indigenous beliefs 0.1%, none 1.7% (2001)

2151. What are the ethnic groups of Rwanda?
Hutu (Bantu) 84%, Tutsi (Hamitic) 15%, Twa (Pygmy) 1%

Saint Helena (United Kingdom)

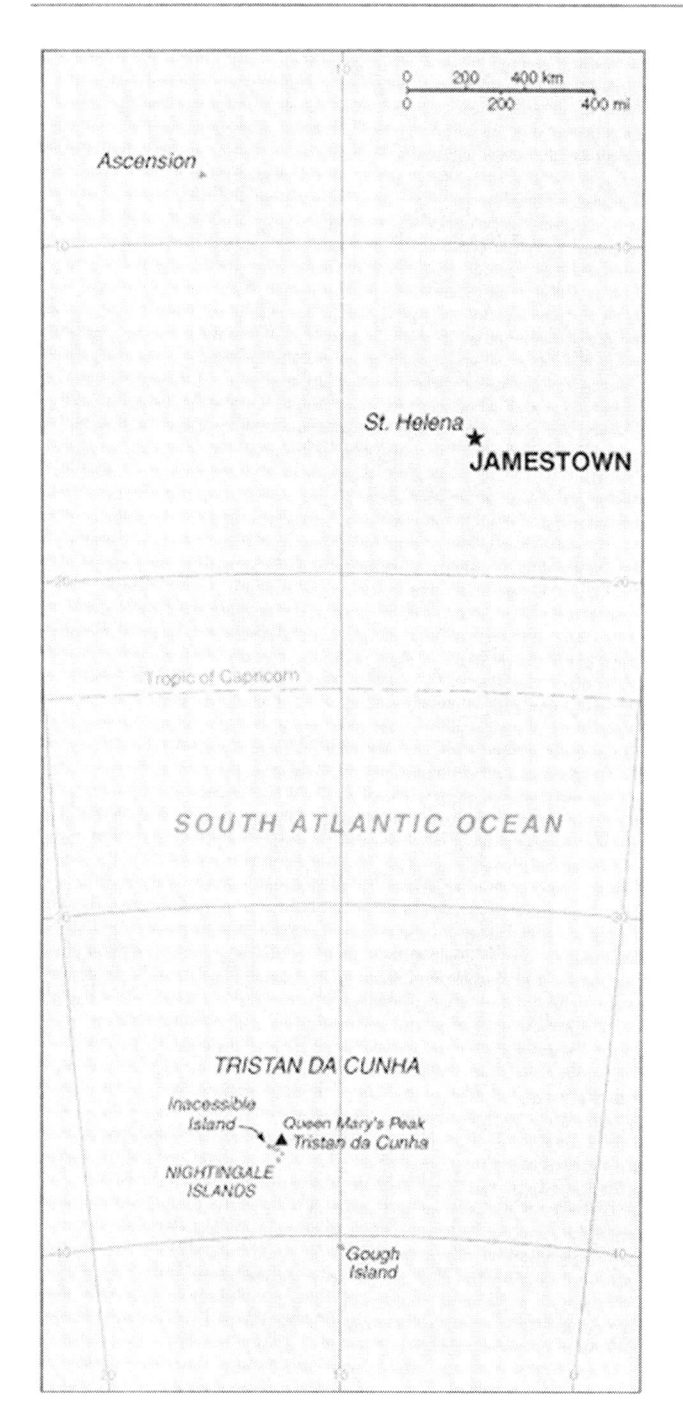

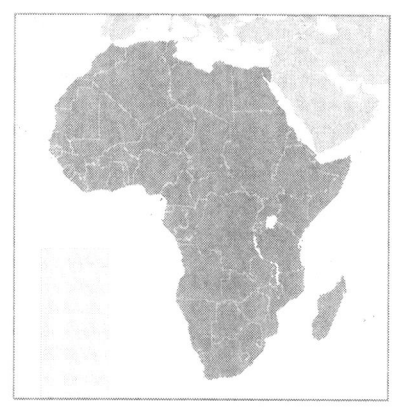

2152. What is Saint Helena?
    A British overseas territory
2153. What is the conventional long form of the territory name?

Saint Helena, Ascension, and Tristan da Cunha

2154. Saint Helena consists of which islands?
Saint Helena and Ascension Islands, and the island group of Tristan da Cunha

2155. Which islands are included in the island group of Tristan da Cunha?
Tristan da Cunha, Nightingale, Inaccessible, and Gough

2156. What is the motto of Saint Helena?
Loyal and Unshakeable

2157. What is the national anthem of Saint Helena?
God Save the Queen

2158. What is the capital of Saint Helena?
Jamestown (the largest city)

2159. What is the official language of Saint Helena?
English

2160. What is the area of Saint Helena?
47 sq mi / 122 km^2

2161. How long is the coastline of Saint Helena Island?
37 mi / 60 km

2162. How long is the coastline of Tristan da Cunha Island?
25 mi / 40 km

2163. What is the population of Saint Helena?
7,637 (by 2009)

2164. What is the currency of Saint Helena?
Saint Helena Pound

2165. What is the geographical feature of the terrain of Saint Helena Island?
Rugged, volcanic; small scattered plateaus and plains

2166. What is the geographical feature of the terrain of Ascension Island?
Surface covered by lava flows and cinder cones of 44 dormant volcanoes; ground rises to the east

2167. What is the geographical feature of the terrain of Tristan da Cunha Island?
Sheer cliffs line the coastline of the nearly circular island; the flanks of the central volcanic peak are deeply dissected; narrow coastal plain lies between The Peak and the coastal cliffs

2168. What is the highest point of Saint Helena Island?
Mount Actaeon (2,684 ft / 818 m)

2169. What is the highest point of Ascension Island?
Green Mountain (2,815 ft / 859 m)

2170. What is the highest point of Tristan da Cunha Island?
Queen Mary's Peak (6,765 ft / 2,062 m)

2171. What is the rank of Queen Mary's Peak by height in South Atlantic Ocean islands?
1st

2172. What is the lowest point of Saint Helena?
Atlantic Ocean (0 ft / 0 m)

2173. How many plant species in Saint Helena are unknown anywhere else in the world?
At least 40

2174. Ascension is a breeding ground for what sea animals?
Sea turtles and sooty terns

2175. What are administrative areas in Saint Helena?
Ascension, Saint Helena, Tristan da Cunha
2176. Which famous leaders were exiled to Saint Helena?
Napoleon Bonaparte (the military and political leader of France and Emperor of the French), Dinuzulu kaCetshwayo (the king of the Zulu nation)
2177. What is the climate of Saint Helena Island?
Tropical marine; mild, tempered by trade winds
2178. What is the climate of Ascension Island?
Tropical marine; mild, semi-arid
2179. What is the climate of Tristan da Cunha Island?
Temperate marine; mild, tempered by trade winds
2180. What are the natural resources of Saint Helena?
Fish, lobster
2181. What are the natural hazards of Saint Helena?
Active volcanism on Tristan da Cunha, last eruption in 1961
2182. What are the religions of Saint Helena?
Anglican (majority), Baptist, Seventh-Day Adventist, Roman Catholic
2183. What are the ethnic groups of Saint Helena?
African descent 50%, white 25%, Chinese 25%

São Tomé and Príncipe

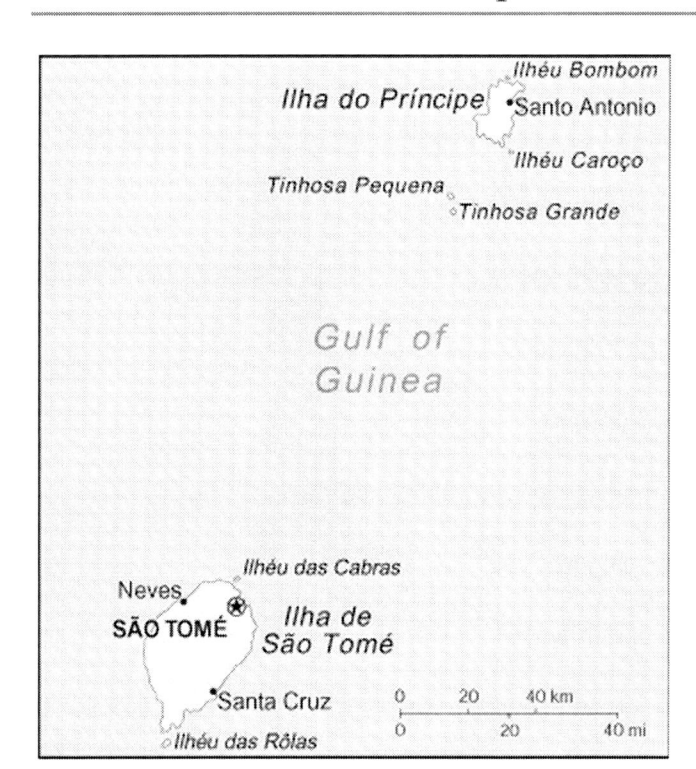

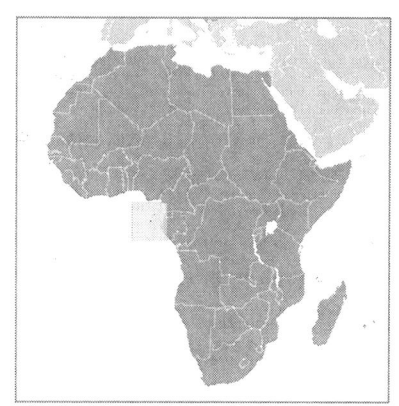

2184. What is the official name of São Tomé and Príncipe?
Democratic Republic of Sao Tome and Principe

2185. São Tomé and Príncipe is located in which gulf?
Gulf of Guinea
2186. What is the distance between Sao Tome Island and Principe Island?
87 mi / 140 km
2187. São Tomé and Príncipe is off the northwestern coast of which country?
Gabon
2188. São Tomé and Príncipe is a part of what geologic rift zone?
The Cameroon Line
2189. What is the national anthem of São Tomé and Príncipe?
Total independence
2190. What is the capital of São Tomé and Príncipe?
São Tomé (the largest city)
2191. What is the official language of São Tomé and Príncipe?
Portuguese
2192. What kind of government does São Tomé and Príncipe have?
Democratic semi-presidential Republic
2193. São Tomé was uninhabited until the arrival of the Portuguese on which day?
December 21st, 1471
2194. Príncipe was uninhabited until the arrival of the Portuguese on which day?
January 17th, 1472
2195. When did São Tomé and Príncipe gain independence from Portugal?
July 12th, 1975
2196. What is the area of São Tomé and Príncipe?
372 sq mi / 1,001 km^2
2197. What is the rank of São Tomé and Príncipe by size in Africa?
Smallest
2198. What is the rank of São Tomé and Príncipe by population in Africa?
Second smallest
2199. How long is the coastline of São Tomé and Príncipe?
130 mi / 209 km
2200. What is the population of São Tomé and Príncipe?
212,679 (by 2009)
2201. What is the currency of São Tomé and Príncipe?
Dobra
2202. What is the geographical feature of the terrain of São Tomé and Príncipe?
Volcanic, mountainous
2203. What is the highest point of São Tomé and Príncipe?
Pico de Sao Tome (6,640 ft / 2,024 m)
2204. What is the lowest point of São Tomé and Príncipe?
Atlantic Ocean (0 ft / 0 m)
2205. Obo National Park covers about what percentage of the surface of São Tomé and Príncipe?
30%
2206. How large is Obo National Park, home to the four different biotopes, the lowland and mountain forests, mangroves and a savanna area?

The Park consists of two areas, one on São Tomé is 91 sq mi / 235 km², and another on Príncipe is 25 sq mi / 65 km²

2207. What are administrative divisions called in São Tomé and Príncipe?
Provinces
2208. How many provinces does São Tomé and Príncipe have?
2
2209. What is the climate of São Tomé and Príncipe?
Tropical; hot, humid; one rainy season (October to May)
2210. What are the natural resources of São Tomé and Príncipe?
Fish, hydropower
2211. What are the religions of São Tomé and Príncipe?
Catholic 70.3%, Evangelical 3.4%, New Apostolic 2%, Adventist 1.8%, other 3.1%, none 19.4%
2212. What are the ethnic groups of São Tomé and Príncipe?
Mestico, angolares (descendants of Angolan slaves), forros (descendants of freed slaves), servicais (contract laborers from Angola, Mozambique, and Cape Verde), tongas (children of servicais born on the islands), Europeans (primarily Portuguese)

Senegal

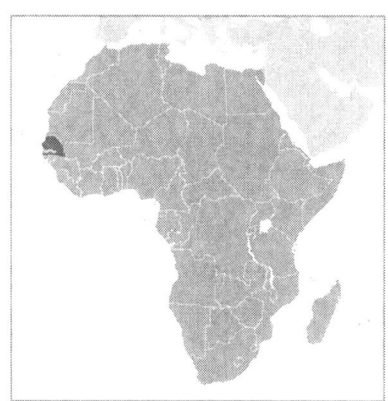

2213. What is the official name of Senegal?
Republic of Senegal
2214. Senegal is south of which river?
Senegal River

2215. Which country borders Senegal to the north?
Mauritania
2216. Which country borders Senegal to the east?
Mali
2217. Which countries border Senegal to the south?
Guinea and Guinea-Bissau
2218. Senegal conclaves which country?
The Gambia
2219. Which water body borders Senegal to the east?
Atlantic Ocean
2220. Is Cap-Vert Peninsula part of Cape Verde Islands?
No, it is a peninsula in Senegal's Atlantic coast, which is 350 miles (560 km) further east of Cape Verde Islands
2221. Is Senegal part of Sahel?
Yes
2222. What is the motto of Senegal?
One People, One Goal, One Faith
2223. What is the national anthem of Senegal?
Everyone strum your koras, strike the balafons
2224. What is the capital of Senegal?
Dakar (the largest city)
2225. What is the official language of Senegal?
French
2226. What kind of government does Senegal have?
Semi-presidential republic
2227. Which part of Senegal belonged to the Ghana Empire (also called Wagadou Empire) during 790–1076?
Eastern part
2228. On April 4th, 1959, Senegal and the French Sudan merged to form which country?
Mali Federation
2229. What was the other country in Mali Federation?
Mali
2230. When did Senegal gain independence from France?
April 4th, 1960
2231. When did Mali Federation become fully independent?
June 20th, 1960
2232. What was Senegambia Confederation?
A loose confederation between Senegal and the Gambia
2233. During which period did Senegambia Confederation exist?
February 1st, 1982–September 30th, 1989
2234. What is the area of Senegal?
76,000 sq mi / 196,723 km^2
2235. How long is the coastline of Senegal?
330 mi / 531 km
2236. What is the population of Senegal?

13,711,597 (by 2009)
2237. What is the currency of Senegal?
CFA franc
2238. What is the geographical feature of the terrain of Senegal?
Generally low, rolling, plains rising to foothills in southeast
2239. What is the highest point of Senegal?
Unnamed feature near Nepen Diakha (1,906 ft / 581 m)
2240. What is the lowest point of Senegal?
Atlantic Ocean (0 ft / 0 m)
2241. Senegal contains what national parks?
Basse Casamance National Park, Isles des Madeleines National Park, Langue de Barbarie National Park, Djoudj National Bird Sanctuary, Niokolo-Koba National Park, and Saloum Delta National Park
2242. What are administrative divisions called in Senegal?
Regions
2243. How many regions does Senegal have?
14
2244. What is the climate of Senegal?
Tropical; hot, humid; rainy season (May to November) has strong southeast winds; dry season (December to April) dominated by hot, dry, harmattan wind
2245. What are the natural resources of Senegal?
Fish, phosphates, iron ore
2246. What are the natural hazards of Senegal?
Lowlands seasonally flooded; periodic droughts
2247. What are the religions of Senegal?
Muslim 94%, Christian 5% (mostly Roman Catholic), indigenous beliefs 1%
2248. What are the ethnic groups of Senegal?
Wolof 43.3%, Pular 23.8%, Serer 14.7%, Jola 3.7%, Mandinka 3%, Soninke 1.1%, European and Lebanese 1%, other 9.4%

Seychelles

2249. What is the official name of Seychelles?
Republic of Seychelles
2250. Zanzibar Island of which country is to the west of Seychelles?
Tanzania
2251. Mauritius and Réunion are located in which direction of Seychelles?
South
2252. Comoros and Mayotte are located in which direction of Seychelles?
Southwest
2253. What is Seychelles' rank by population in Africa?
Smallest
2254. How many islands are there in Seychelles?
155 (listed by Constitution of the Republic of Seychelles, listed by CIA–41 granitic islands and 75 coralline islands)

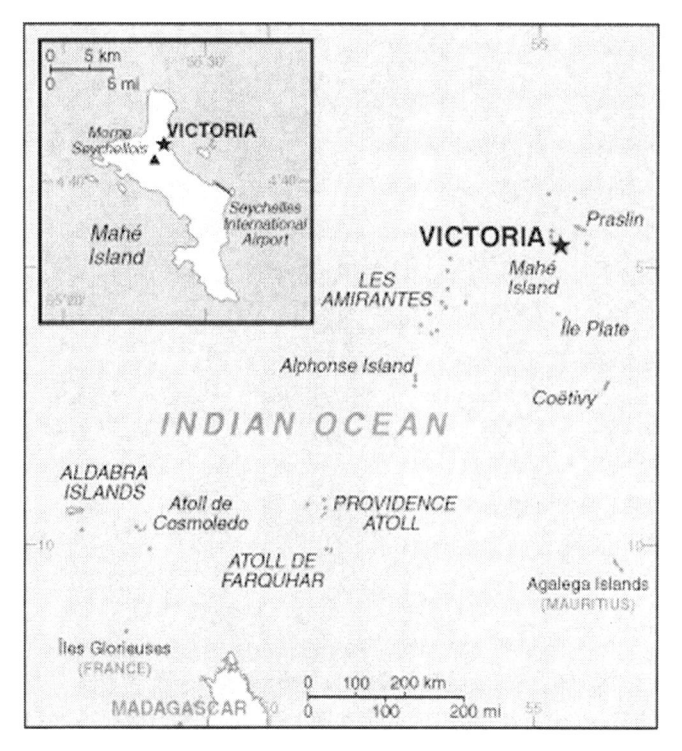

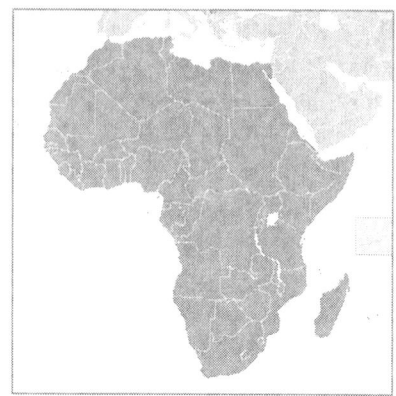

2255. What is the motto of Seychelles?
The End Crowns the Work
2256. What is the national anthem of Seychelles?
Join together all Seychellois
2257. What is the capital of Seychelles?
Victoria (the largest city)
2258. What is the biggest island on which Victoria is located?
Mahé (55 sq mi / 142 km^2)
2259. What are the official languages of Seychelles?
Seychellois Creole, English and French
2260. What kind of government does Seychelles have?
Republic
2261. When did French take control of Seychelles?
1756
2262. Which country contested control over Seychelles with the French between 1794 and 1810?
United Kingdom
2263. When did United Kingdom take control of Seychelles and Mauritius?
1810
2264. When did United Kingdom formalize the control of Seychelles by Treaty of Paris?
1814
2265. When did Seychelles become a crown colony of United Kingdom, separating from Mauritius?
1903
2266. When did Seychelles gain independence from United Kingdom?

June 29th, 1976
2267. What is the area of Seychelles?
174 sq mi / 451 km^2
2268. How long is the coastline of Seychelles?
305 mi / 491 km
2269. What is the population of Seychelles?
87,476 (by 2009)
2270. What is the currency of Seychelles?
Seychellois Rupee
2271. What is the geographical feature of the terrain of Seychelles?
Mahe Group is granitic, narrow coastal strip, rocky, hilly; others are coral, flat, elevated reefs
2272. What is the highest point of Seychelles?
Morne Seychellois (2,969 ft / 905 m)
2273. What is the lowest point of Seychelles?
Indian Ocean (0 ft / 0 m)
2274. Seychelles contains what national parks?
Baie Ternay Marine National Park, Curieuse Marine National Park, Ile Coco Marine National Park, Morne Seychellois National Park, Port Launay Marine National Park, Praslin National Park, Silhouette Marine National Park, and Ste Anne Marine National Park
2275. What are administrative divisions called in Seychelles?
Districts
2276. How many districts does Seychelles have?
26
2277. What is the climate of Seychelles?
Tropical marine; humid; cooler season during southeast monsoon (late May to September); warmer season during northwest monsoon (March to May)
2278. What are the natural resources of Seychelles?
Fish, copra, cinnamon trees
2279. What are the natural hazards of Seychelles?
Lies outside the cyclone belt, so severe storms are rare; short droughts possible
2280. What are the religions of Seychelles?
Roman Catholic 82.3%, Anglican 6.4%, Seventh Day Adventist 1.1%, other Christian 3.4%, Hindu 2.1%, Muslim 1.1%, other non-Christian 1.5%, unspecified 1.5%, none 0.6%
2281. What are the ethnic groups of Seychelles?
Mixed French, African, Indian, Chinese, and Arab

Sierra Leone

2282. What is the official name of Sierra Leone?
Republic of Sierra Leone
2283. Which country borders Sierra Leone to the north?
Guinea
2284. Which country borders Sierra Leone to the southeast?
Liberia

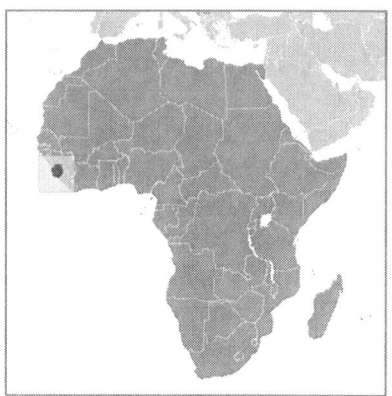

2285. Which water body borders Sierra Leone to the southwest?
Atlantic Ocean
2286. What is the motto of Sierra Leone?
Unity, Freedom, Justice
2287. What is the national anthem of Sierra Leone?
High We Exalt Thee, Realm of the Free
2288. What is the capital of Sierra Leone?
Freetown (the largest city)
2289. What is the official language of Sierra Leone?
English
2290. Which language is used by 97% of Sierra Leone?
Krio
2291. What kind of government does Sierra Leone have?
Constitutional republic
2292. When did United Kingdom declare Sierra Leone to be a protectorate?
August 31st, 1896
2293. What was the Hut Tax War of 1898?
A war of resistance to British colonialism in Sierra Leone
2294. What was the immediate trigger of the Hut Tax War of 1898?
The hut tax imposed on January 1st, 1898
2295. What was the result of the Hut Tax War of 1898?
The native leaders, Bai Bureh, Nyagua and Be Sherbro (Gbana Lewis), were exiled to the Gold Coast on July 30th, 1899; a large number of their subordinates were executed
2296. When did Sierra Leone gain independence from United Kingdom?

April 27th, 1961
2297. When did Sierra Leone Civil War happen?
March 23rd, 1991–January 18th, 2002
2298. How many people died in the Sierra Leone Civil War?
About 100,000
2299. Why did Sierra Leone Civil War happen?
Over two decades of government neglect of the interior followed by the spilling over of the Liberian conflict into its borders
2300. What was the result of Sierra Leone Civil War?
Nigeria led United Nations troops plus British force defeat the rebel forces and restored the civilian government elected
2301. What is the area of Sierra Leone?
27,699 sq mi / 71,740 km^2
2302. How long is the coastline of Sierra Leone?
250 mi / 402 km
2303. What is the population of Sierra Leone?
5,132,138 (by 2009)
2304. What is the currency of Sierra Leone?
Leone
2305. What is the geographical feature of the terrain of Sierra Leone?
Coastal belt of mangrove swamps, wooded hill country, upland plateau, mountains in east
2306. What is the highest point of Sierra Leone?
Loma Mansa (also called Bintumani, 6,391 ft / 1,948 m)
2307. Loma Mansa is located in which mountain range?
Loma Mountains
2308. What is the lowest point of Sierra Leone?
Atlantic Ocean (0 ft / 0 m)
2309. Is Queen Elizabeth II Quay the third largest natural harbor in the world?
No. Although quoted often, it is not true. It is a man-made port and not the third largest
2310. Which river is the largest in the Republic of Sierra Leone?
Rokel River
2311. What kind of river is the Sierra Leone River in Western Sierra Leone?
Estuary
2312. Sierra Leone River is a natural harbour. What is its ranking by size in the African continent?
1st (1,139 sq mi / 2,950 km²)
2313. What is the largest island in Sierra Leone River?
Tasso Island
2314. What is the historically important island in Sierra Leone River?
Bunce Island (a site of an 18th century British slave castle)
2315. Sierra Leone contains what national parks?
Kuru Hills National Park, Lake Mape/Mabesi National Park, Lake Sonfon National Park, Loma Mountains National Park, Outamba-Kilimi National Park, and Western Area National Park
2316. What are administrative divisions called in Sierra Leone?
Provinces

2317. How many provinces does Sierra Leone have?
    3 (plus 1 area)
2318. What is the climate of Sierra Leone?
    Tropical; hot, humid; summer rainy season (May to December); winter dry season (December to April)
2319. What are the natural resources of Sierra Leone?
    Diamonds, titanium ore, bauxite, iron ore, gold, chromite
2320. What are the natural hazards of Sierra Leone?
    Dry, sand-laden harmattan winds blow from the Sahara (December to February); sandstorms, dust storms
2321. What are the religions of Sierra Leone?
    Muslim 60%, Christian 10%, indigenous beliefs 30%
2322. What are the ethnic groups of Sierra Leone?
    20 African ethnic groups 90% (Temne 30%, Mende 30%, other 30%), Creole (Krio) 10% (descendants of freed Jamaican slaves who were settled in the Freetown area in the late-18th century), refugees from Liberia's recent civil war, small numbers of Europeans, Lebanese, Pakistanis, and Indians

Somalia

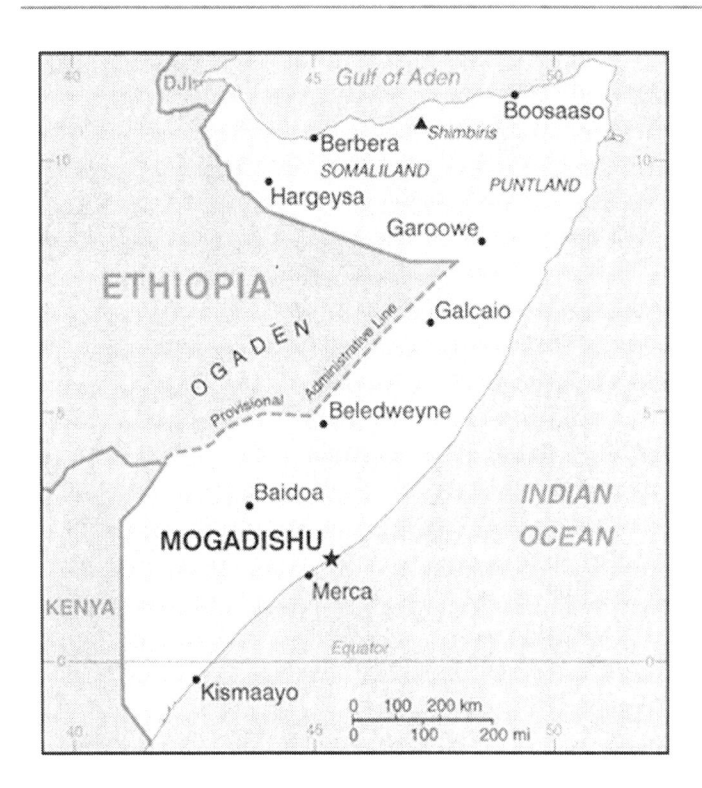

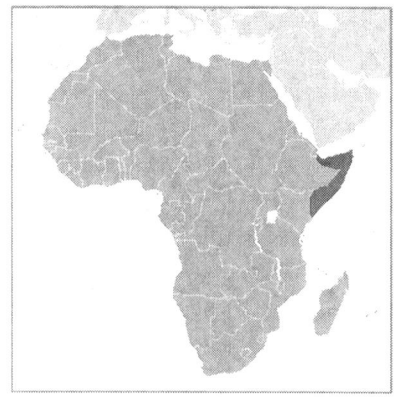

2323. What is the official name of Somalia?
    Republic of Somalia
2324. What country borders Somalia to the northwest?
    Djibouti

2325. What country borders Somalia to the southwest?
   Kenya
2326. Which country borders Somalia to the west?
   Ethiopia
2327. What body of water borders Somalia to the east?
   Indian Ocean
2328. What body of water borders Somalia to the north?
   Gulf of Aden
2329. What is the national anthem of Somalia?
   Somalia, Wake Up
2330. What is the capital of Somalia?
   Mogadishu (the largest city)
2331. What are the official languages of Somalia?
   Somali, Arabic
2332. What kind of government does Somalia have?
   Coalition government (no permanent national government; transitional, parliamentary federal government)
2333. What was Somalia's former name?
   Somali Democratic Republic
2334. When did Somalia gain independence from United Kingdom?
   June 26th, 1960
2335. When did Somalia gain independence from Italy?
   July 1st, 1960
2336. When did the Somali Civil War first start?
   January 26th, 1991 after Siad Barre was ousted from power
2337. The Somali Civil War resulted in the birth of what kind of crime?
   Piracy
2338. When was the First Battle of Mogadishu fought?
   October 3rd, 1993–October 4th, 1993
2339. The First Battle of Mogadishu was fought between which two parties?
   UNOSOM II (US, Malaysian, and Pakistani troops) and Somali National Alliance
2340. What was the result of the First Battle of Mogadishu?
   The forced withdrawal of UNOSOM troops
2341. What was the original intent of United States presence in Somalia during the conflict?
   To provide food and aid for Somalis stuck in war
2342. The civil war in Somalia that started in 2006 was between which two parties?
   Islamic Court Union and Alliance for the Restoration of Peace and Counter-Terrorism
2343. Which side won the Second Battle of Mogadishu?
   Islamic Court Union
2344. What happened to Somalia's government on December 28th, 1993?
   The Fall of Mogadishu
2345. What happened during the Fall of Mogadishu?
   Somalia's Transitional Federal Government (TFG) and Ethiopian troops captured Mogadishu, defeating the Islamic Court Union
2346. How did the Third Battle of Mogadishu start?

Insurgent attacks on TFG and Ethiopian troops
2347. What was the result of the Third Battle of Mogadishu?
An indecisive victory for TFG and Ethiopia and increased insurgency
2348. What happened to provoke another conflict in Mogadishu in November 2007?
The dragging of dead Ethiopian soldiers on the streets of Mogadishu
2349. What is the area of Somalia?
246,201 sq mi / 637,661 km^2
2350. How long is the coastline of Somalia?
1,880 mi / 3,025 km
2351. What is the population of Somalia?
9,133,000 (by 2009)
2352. What is the currency of Somalia?
Somali Shillings
2353. What is the geographical feature of the terrain of Somalia?
Mostly flat to undulating plateau rising to hills in north
2354. What is the highest point of Somalia?
Shimbiris (7,927 ft / 2,416 m)
2355. Shimbiris is part of the Surud Range, a part of what mountain range?
Golis Mountains
2356. What is the lowest point of Somalia?
Indian Ocean (0 ft / 0 m)
2357. The coastal plains of Somalia are a part of which ecoregion?
Ethiopian xeric grasslands and shrublands ecoregion
2358. The northern half of Somalia is known by what name?
Somaliland
2359. When did Somaliland declare its independence from Somalia?
May 18th, 1991
2360. Somaliland's independence is recognized by how many countries?
0
2361. What is Somaliland's motto?
Justice, Peace, Freedom, Democracy and Success for All
2362. What is the national anthem of Somaliland?
Sama ku waar
2363. What is the currency of Somaliland?
Somaliland Shilling
2364. What is the capital of Somaliland?
Hargesia
2365. Somalia is at the tip of what region of Africa?
The Horn of Africa
2366. What is the region containing the tip of the Horn of Africa called?
Puntland
2367. When did Puntland declare its independence from Somalia?
1998
2368. Puntland's independence is recognized by how many countries?
0

2369. What is the official name of Puntland?
Puntland State of Somalia
2370. What is the capital of Puntland?
Garowe
2371. What is the origin of the name Puntland?
From the Land of Punt, a trading partner of ancient Egypt
2372. How many permanent rivers are there in Somalia?
2 (Jubba River and Shabele River)
2373. Which mammals are endemic to Somalia?
The Silver Dik-dik and the Somali Golden Mole
2374. Somalia contains what national parks?
Kismayu National Park and Hargeysa National Park
2375. What are administrative divisions called in Somalia?
Regions
2376. How many districts does Somalia have?
18
2377. What is the climate of Somalia?
Principally desert; northeast monsoon (December to February), moderate temperatures in north and hot in south; southwest monsoon (May to October), torrid in the north and hot in the south, irregular rainfall, hot and humid periods (tangambili) between monsoons
2378. What are the natural resources of Somalia?
Uranium and largely unexploited reserves of iron ore, tin, gypsum, bauxite, copper, salt, natural gas, likely oil reserves
2379. What are the natural hazards of Somalia?
Recurring droughts; frequent dust storms over eastern plains in summer; floods during rainy season
2380. What are the religions of Somalia?
Sunni Muslim
2381. What are the ethnic groups of Somalia?
Somali 85%, Bantu and other non-Somali 15% (including Arabs 30,000)

South Africa

2382. What is the official name of South Africa?
Republic of South Africa
2383. Which countries border South Africa to the north?
Namibia, Botswana and Zimbabwe
2384. Which countries border South Africa to the east?
Mozambique and Swaziland
2385. Which country was landlocked by South Africa?
Lesotho
2386. What is the motto of South Africa?
Unity In Diversity
2387. What is the national anthem of South Africa?
National anthem of South Africa

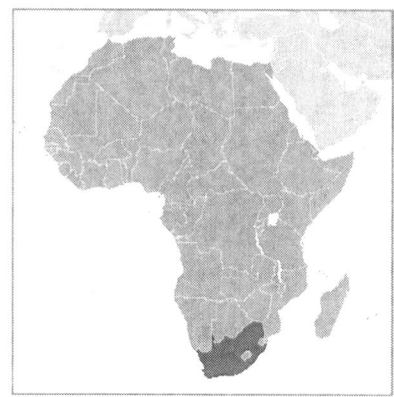

2388. What is the administrative capital of South Africa?
Pretoria
2389. What is the legislative capital of South Africa?
Cape Town
2390. What is the judicial capital of South Africa?
Bloemfontein
2391. What is the largest city of South Africa?
Johannesburg
2392. What was Witwatersrand Gold Rush?
The gold rush in 1886 that led to the establishment of Johannesburg
2393. What are official languages of South Africa?
Afrikaans, English, Southern Ndebele, Northern Sotho, Southern Sotho, Swazi, Tsonga, Tswana, Venda, Xhosa, and Zulu
2394. What kind of government does South Africa have?
Constitutional democracy
2395. When did Kingdom of Mapungubwe exist?
975–1220
2396. What was the capital of Kingdom of Mapungubwe?
Mapungubwe
2397. What was the language of Kingdom of Mapungubwe?
Shona-iKalanga
2398. What was the religion of Kingdom of Mapungubwe?
Cult of Mwari
2399. When did Republic of Swellendam exist?

1795
2400. What was the significance that Hermanus Steyn was appointed as the president of Republic of Swellendam?
Hermanus Steyn was the first to be appointed the title of President in South Africa
2401. When did British colony Cape Colony, along the southern coast of the present-day South Africa, exist?
1795–1910
2402. What was the capital of Cape Colony?
Cape Town
2403. What languages were used in Cape Colony?
English and Dutch
2404. What was the area of Cape Colony by 1910?
219,700 sq mi / 569,020 km^2
2405. What was the population of Cape Colony by 1910?
2,564,965
2406. What was the currency of Cape Colony?
Pound Sterling
2407. When was Natalia Republic, on the coast of the Indian Ocean beyond the Eastern Cape, established?
October 12th, 1839
2408. What was the capital of Natalia Republic?
Pietermaritzburg
2409. What languages were used in Natalia Republic?
Dutch, Zulu, English
2410. When did Battle of Blood River happen?
December 16th, 1838
2411. Which river was red during Battle of Blood River?
Ncome River
2412. Who fought Zulu Kingdom during Battle of Blood River?
Voortrekkers, emigrants during the 1830s and 1840s who left the Cape Colony and moving into the interior
2413. When did Natalia Republic ally with Zulu Kingdom?
January 1840
2414. When was Natalia Republic annexed by United Kingdom?
May 12th, 1843
2415. When did Colony of Natal exist?
1843–1910
2416. When did Zulu Kingdom, in eastern plateau of present-day South Africa, exist?
1816–1897
2417. What was the capital of Zulu Kingdom?
KwaBulawayo (Zimbabwe), later Ulundi
2418. What was the area of Zulu Kingdom by 1828?
11,500 sq mi / 29,785 km^2
2419. What was the population of Zulu Kingdom by 1828?
250,000

2420. What was the currency of Zulu Kingdom?
Cattle
2421. What was Zulu Civil War?
A war fought between the expanding Zulu Kingdom and the Ndwandwe tribe in South Africa
2422. When did Zulu Civil War (also called Ndwandwe-Zulu War) happen?
1817–1819
2423. When did Battle of Gqokli Hill, a part of the Zulu Civil War, happened?
1818
2424. When did Battle of Mhlatuze River, a battle following the Zulu Civil War, happened?
1820
2425. What was Anglo-Zulu War?
A war fought between United Kingdom and Zulu Kingdom
2426. During which period did Anglo-Zulu War happen?
January 11th, 1879–July 4th, 1879
2427. What was the result of Anglo-Zulu War?
United Kingdom annexed Zulu Kingdom
2428. When was South African Republic (also called Transvaal Republic) established?
June 27th, 1856
2429. What was national anthem of South African Republic?
National anthem of the Transvaal
2430. What was the capital of South African Republic?
Pretoria
2431. What languages were used in South African Republic?
Dutch (written), Afrikaans (spoken)
2432. What currency was used in South African Republic?
South African Pound
2433. When was South African Republic annexed by United Kingdom?
1877
2434. During which period did First Boer War happen?
December 29th, 1880 - March 23rd, 1881
2435. What was the result of First Boer War?
South African Republic regained independence in 1881
2436. What was Second Boer War (also called the Boer War, South African War the Anglo-Boer War or Engelse oorlog)?
The war between United Kingdom and the two independent Boer republics, South African Republic and the Orange Free State
2437. During which period did Second Boer War happen?
October 11th, 1899–May 31st, 1902
2438. What was signed after Second Boer War?
Treaty of Vereeniging (also called Peace of Vereeniging) was signed on May 31st, 1902
2439. According to Treaty of Vereeniging, what happened to South African Republic?
It became Transvaal Colony
2440. When was Republic of the Orange Free State, the present-day Free State province in South Africa, established?

February 17th, 1854
2441. What was the capital of Republic of the Orange Free State?
Bloemfontein
2442. Which languages were used in Republic of the Orange Free State?
Dutch, Afrikaans, English, Sesotho, and Zulu
2443. Which war did Republic of the Orange Free State fought, First Boer War or Second Boer War?
Second Boer War
2444. According to Treaty of Vereeniging, what happened to Republic of the Orange Free State?
It became Orange River Colony
2445. When was Union of South Africa established?
May 31st, 1910
2446. Which colonies were preceded to Union of South Africa?
Cape Colony, Colony of Natal, Orange River Colony, Transvaal Colony, and German South-West Africa Colony
2447. What kind of government was Union of South Africa?
It was founded as a dominion, later Commonwealth realm
2448. What was motto of Union of South Africa?
Unity, strength
2449. When did Union of South Africa become Republic of South Africa?
May 31st, 1961
2450. What is the area of South Africa?
471,443 sq mi / 1,221,037 km^2
2451. How long is the coastline of South Africa?
1,739 mi / 2,798 km
2452. What is the population of South Africa?
49,052,489 (by 2009)
2453. What is the currency of South Africa?
Rand
2454. What is the geographical feature of the terrain of South Africa?
Vast interior plateau rimmed by rugged hills and narrow coastal plain
2455. What is the highest point of South Africa?
Njesuthi (11,181 ft / 3,408 m)
2456. Njesuthi is located in which mountain range?
Drakensberg
2457. What is the lowest point of South Africa?
Atlantic Ocean (0 ft / 0 m)
2458. What was the longest river in South Africa?
Orange River
2459. What is the largest tributary of Orange River in South Africa?
Vaal River (696 mi / 1,120 km)
2460. What is the largest lake in South Africa?
Lake Chrissie (6 miles / 9 km long and 2 miles / 3 km wide and has a circumference of 14 miles / 25 km)
2461. What are Prince Edward Islands?

Two small islands in the sub-antarctic Indian Ocean that are politically part of South Africa
2462. What are two island names of Prince Edward Islands?
Marion Island (110 sq mi / 290 km^2) and Prince Edward Island (17 sq mi / 45 km^2)
2463. What is Karoo which occupies a large part of the western Central Plateau?
A semi-desert region
2464. What are the two main sub-regions in Karoo?
The Great Karoo in the north and the Little Karoo in the south
2465. What is Garden Route in South Africa?
A popular and scenic stretch of the southeastern coast of South Africa
2466. What is Cape of Good Hope?
A rocky headland on the Atlantic coast of South Africa
2467. Is Cape of Good Hope the southernmost tip of Africa?
No, the southernmost point is Cape Agulhas
2468. What is Cape Fold Belt?
The folded sedimentary sequence of rocks in the southwestern corner of South Africa
2469. What is Albany Thickets?
An ecoregion of dense woodland near the southern point of South Africa, mostly in the Eastern Cape
2470. Kalahari Desert covers South Africa and which other two countries?
Botswana and Namibia
2471. Witwatersrand, a low, sedimentary range of hills, is famous for what?
Being the source of 40% of the gold ever mined from the earth
2472. What is the smallest national park in South Africa?
Bontebok National Park (11 sq mi / 28 km^2)
2473. Bontebok National Park, located in Western Cape of South Africa, was established to ensure the preservation of which animal?
Bontebok
2474. How large is Addo Elephant National Park, located close to Port Elizabeth in South Africa?
630 sq mi / 1,640 km^2
2475. How large is Agulhas National Park, located in Agulhas Plain in the southern Overberg region of the Western Cape?
81 sq mi / 469 km^2
2476. What is Agulhas National Park famous for?
Cape Agulhas, the southernmost tip of Africa
2477. How large is Augrabies Falls National Park, located in Northern Cape Province of South Africa?
320 sq mi / 820 km^2
2478. What is Augrabies Falls National Park famous for?
Augrabies Falls
2479. How large is Camdeboo National Park, located in Karoo of South Africa?
75 sq mi / 194 km^2
2480. Which national park was established on March 6th, 2009 by amalgamating the existing Tsitsikamma and Wilderness National Parks, the Knysna National Lake Area, and various other areas of state-owned land?
Garden Route National Park

2481. What was Tsitsikamma National Park famous for?
Indigenous forests, dramatic coastline, and the Otter Trail
2482. How large is Garden Route National Park, which is famous for Knysna-Amatole montane forests?
470 sq mi / 1,210 km^2
2483. How large is Golden Gate Highlands National Park, located in Free State of South Africa?
131 sq mi / 340 km²
2484. What is Golden Gate Highlands National Park famous for?
Golden, ochre, and orange-hued deeply eroded sandstone cliffs and outcrops, especially the Brandwag rock
2485. How large is Karoo National Park, a wildlife reserve in the isolated Karoo area of the Western Cape?
296 sq mi / 768 km^2
2486. Kgalagadi Transfrontier Park, the wildlife preserve and conservation area between South Africa and Botswana, includes which park in South Africa?
Kalahari Gemsbok National Park
2487. How large is Kruger National Park in the east of South Africa, which is home to many species: 336 trees, 49 fish, 34 amphibians, 114 reptiles, 507 birds and 147 mammals?
7,332 sq mi / 18,989 km^2
2488. How large is Mapungubwe National Park, located in Limpopo Province of South Africa?
110 sq mi / 280 km^2
2489. What is Mapungubwe National Park famous for?
The historical site of Mapungubwe Hill, which was the capital of Kingdom of Mapungubwe
2490. How large is Marakele National Park (also called Kransberg National Park), located in Limpopo Province of South Africa?
260 sq mi / 670 km^2
2491. Marakele is home of what animals?
The "big five" (lion, leopard, buffalo, rhinoceros and elephant) as well as sixteen species of antelopes and 250 species of birds
2492. How large is Mokala National Park, located in Northern Cape of South Africa?
77 sq mi / 200 km^2
2493. Mokala National Park effectively replaces which national park that was deproclaimed to comply with diamond prospecting rights?
Vaalbos National Park
2494. How large is Mountain Zebra National Park, located in Eastern Cape of South Africa?
77 sq mi / 200 km²
2495. Mountain Zebra National Park was proclaimed for the nature reserve for which endangered animal?
Cape mountain zebra
2496. How large is Namaqua National Park, located in Northern Cape of South Africa?
270 sq mi / 700 km^2
2497. What is Namaqua National Park famous for?
Abundant spring bloom of brightly colored wildflowers
2498. How large is Table Mountain National Park (also called Cape Peninsula National Park), located in Cape Town of South Africa?

85 sq mi / 221 km^2
2499. What are two landmarks of Table Mountain National Park?
Table Mountain and Cape of Good Hope
2500. How large is Tankwa Karoo National Park, located near the border of Northern Cape and Western Cape in South Africa, which is home to a large variety of birds?
430 sq mi / 1,110 km^2
2501. How large is West Coast National Park, located in Western Cape of South Africa, which surrounds Langebaan Lagoon?
106 sq mi / 275 km^2
2502. What is West Coast National Park famous for?
Its bird life and the spring flowers
2503. Ai-Ais/Richtersveld Transfrontier Park is a peace park straddling the border between South Africa and Namibia. Which national park in South African is part of Ai-Ais/Richtersveld Transfrontier Park?
Richtersveld National Park
2504. What are administrative divisions called in South Africa?
Provinces
2505. How many provinces does South Africa have?
9
2506. What is the climate of South Africa?
Mostly semiarid; subtropical along east coast; sunny days, cool nights
2507. What are the natural resources of South Africa?
Gold, chromium, antimony, coal, iron ore, manganese, nickel, phosphates, tin, uranium, gem diamonds, platinum, copper, vanadium, salt, natural gas
2508. What are the natural hazards of South Africa?
Prolonged droughts
2509. What are the religions of South Africa?
Zion Christian 11.1%, Pentecostal/Charismatic 8.2%, Catholic 7.1%, Methodist 6.8%, Dutch Reformed 6.7%, Anglican 3.8%, Muslim 1.5%, other Christian 36%, other 2.3%, unspecified 1.4%, none 15.1%
2510. What are the ethnic groups of South Africa?
Black African 79%, white 9.6%, colored 8.9%, Indian/Asian 2.5%

Sudan

2511. What is the official name of Sudan?
Republic of Sudan
2512. Which countries border Sudan to the west?
Chad and Central African Republic
2513. Which country borders Sudan to the north?
Egypt
2514. What countries border Sudan to the east?
Eritrea and Ethiopia
2515. What country borders Sudan to the southeast?
Kenya

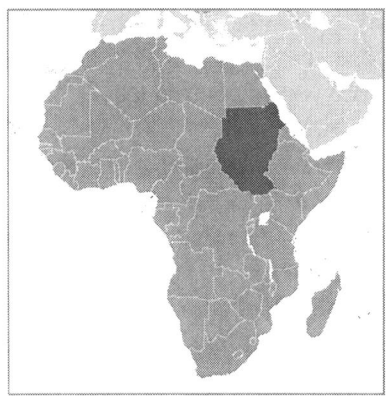

2516. What country borders Sudan to the south?
Uganda
2517. What country borders Sudan to the southwest?
Democratic Republic of the Congo
2518. What country borders Sudan to the northwest?
Libya
2519. What is the motto of Sudan?
Victory is Ours
2520. What is the national anthem of Sudan?
We are the Army of God and Our Land
2521. What is the capital of Sudan?
Khartoum
2522. What is the largest city of Sudan?
Omdurman
2523. What are the official languages of Sudan?
Arabic and English
2524. What kind of government does Sudan have?
Federal Presidential Democratic Republic
2525. Sudan's history is very closely entwined with which other country?
Egypt
2526. Archeological evidence found in Sudan shows that humans have lived there for at least how many years?
60,000
2527. What kingdom ruled present day Sudan during 600 BCE-350 CE?

Meroitic Kingdom
2528. The Meroitic Kingdom fell in 350 CE to which kingdoms?
Abyssinia
2529. Which kingdoms were successors of the Meroitic Kingdom?
Nobatia, Alawa and Maqurra
2530. At 540 CE, the kingdoms switched to which form of religion?
Christianity
2531. What religion did the kingdoms switch to later in the millennium?
Islam
2532. In 1820, which Albanian Ottoman ruler attacked and conquered Sudan
Muhammad Ali Pasha
2533. Who was appointed leader of Sudan under Muhammad Ali Pasha?
Ismail I
2534. In 1879, European powers forced Ismail I to step down to give power to whom?
Tefwik I
2535. Tefwik I's rise to power caused what rebellion to take place?
Urabi Revolt
2536. In 1885, Muhammed Ahmad ibn Abd Allah rebelled and overthrew the Khedivial government for what?
The Mahdiyah (Mahdist Regime)
2537. How was the Mahdiyah run?
Like a military camp
2538. In 1899, United Kingdom invaded Sudan under the name of which government?
Egyptian Khedive
2539. The United Kingdom ruled Sudan as which two separate entities?
The Muslim North and the Christian South
2540. When did Sudan gain independence from United Kingdom and Egypt?
January 1st, 1956
2541. Which war soon started in 1955?
First Sudanese Civil War
2542. Which parties fought each other during the First Sudanese Civil War?
Christians (South) and Muslims (North)
2543. The insurgent party (South) organized under what name?
Anyanya
2544. As a result of the war, what happened to the official government of Sudan (North)?
The government was overthrown by a coupe d'état several times
2545. The war continued to 1971, when Joseph Lagu combined the guerilla armies under which new organization?
Southern Sudan Liberation Movement (SSLM)
2546. The First Sudanese Civil War finally ended on March 1972 with what document?
Addis Ababa Agreement
2547. Conflict restarted with the Second Sudanese Civil War on which year?
1983
2548. Which two sides fought in the Second Sudanese Civil War?
Sudan/ Lord's Resistance Army against Sudan's Liberation Army Front/Eastern Front

2549. On which year did senior military leaders name Sudan an Islamic State?
1985
2550. The Second Sudanese Civil War was revived despite elections that promised peace because of what event?
Military takeover of the government on June 30th, 1989
2551. During 2002, USA issued the Sudan Peace Act, which accused Sudan's government of what crime?
Genocide
2552. Peace talks eventually led to what treaty, which was signed on January 9th, 2005?
Comprehensive Peace Agreement
2553. What was the total death toll of the Second Sudanese Civil War?
1.9 million
2554. Continued acts of genocide still occur today, with most of it happening in which region of Sudan?
Darfur
2555. The Nile River, the longest river in the world, is formed with the union of the Blue and White Niles in which Sudanese city?
Khartoum
2556. The Blue Nile forms one of the largest wetlands in the world. What is the wetlands' name?
As Sudd (also called The Sudd, Bahr el Jebel, or Al Sudd)
2557. What is the area of Sudan?
967,495 sq mi / 2,505,813 km^2
2558. How long is the coastline of Sudan?
530 mi / 853 km
2559. What is the population of Sudan?
41,087,825 (by 2009)
2560. What is the currency of Sudan?
Sudanese Pound
2561. What is the geographical feature of the terrain of Sudan?
Generally flat, featureless plain; mountains in far south, northeast and west; desert dominates the north
2562. What is the highest point of Sudan?
Kinyeti (10,456 ft / 3,187 m)
2563. What is the lowest point of Sudan?
Red Sea (0 ft / 0 m)
2564. Sudan contains what national parks?
Dinder National Park, Radom National Park, Royal Natal National Park, and Southern National Park
2565. What are administrative divisions called in Sudan?
States
2566. How many states does Sudan have?
25
2567. What is the climate of Sudan?
Tropical in south; arid desert in north; rainy season varies by region (April to November)
2568. What are the natural resources of Sudan?

Petroleum; small reserves of iron ore, copper, chromium ore, zinc, tungsten, mica, silver, gold, hydropower
2569. What are the natural hazards of Sudan?
Dust storms and periodic persistent droughts
2570. What are the religions of Sudan?
Sunni Muslim 70% (in north), Christian 5% (mostly in south and Khartoum), indigenous beliefs 25%
2571. What are the ethnic groups of Sudan?
Black 52%, Arab 39%, Beja 6%, foreigners 2%, other 1%

Swaziland

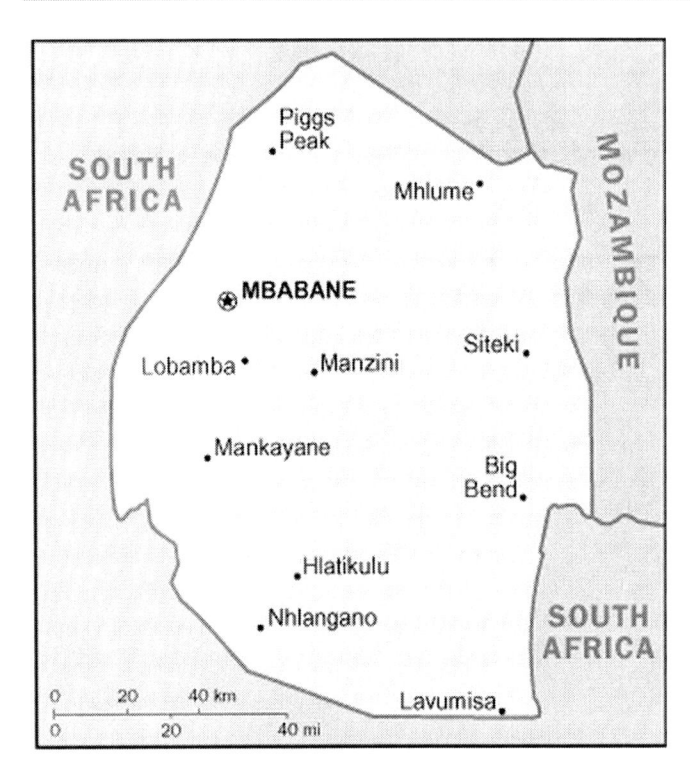

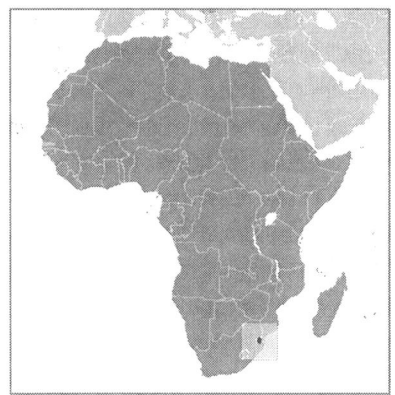

2572. What is the official name of Swaziland?
Kingdom of Swaziland
2573. Which country borders Swaziland to the north, the west, and the south?
South Africa
2574. Which country borders Swaziland to the east?
Mozambique
2575. What are mottos of Swaziland?
Siyinqaba (SiSwati), We are a fortress, We are a mystery/riddle, We hide ourselves away
2576. What is the national anthem of Swaziland?
O Lord our God, bestower of the blessings of the Swazi
2577. What is the royal and legislative capital of Swaziland?
Lobamba

2578. What is the administrative capital of Swaziland?
Mbabane (the largest city)
2579. What are official languages of Swaziland?
English and SiSwati
2580. What kind of government does Swaziland have?
Absolute Monarchy
2581. After Anglo-Boer War, British forces entered Swaziland in 1902, proclaiming British overrule and jurisdiction in 1903, initially as part of which colony of South Africa?
Transvaal Colony
2582. In which year was Swaziland separated from Transvaal Colony?
1906
2583. When did Swaziland gain independence from United Kingdom?
September 6th, 1968
2584. What is the area of Swaziland?
6,704 sq mi / 17,364 km^2
2585. How long is the coastline of Swaziland?
0 mi / 0 km
2586. What is the population of Swaziland?
1,337,186 (by 2009)
2587. What is the currency of Swaziland?
Lilangeni
2588. What is the geographical feature of the terrain of Swaziland?
Mostly mountains and hills; some moderately sloping plains
2589. What is the highest point of Swaziland?
Emlembe (6,109 ft / 1,862 m)
2590. Emlembe is located in which mountain range?
uKhahlamba (also called Drakensberg)
2591. What is the lowest point of Swaziland?
Great Usutu River (69 ft / 21 m)
2592. Great Usutu River (also called Maputo River, Lusutfu River or Suthu River) is a river in Swaziland and which other two countries?
South Africa and Mozambique
2593. What is Swaziland's largest park, which was a private royal hunting ground?
Hlane Royal National Park (116 sq mi / 300 km^2)
2594. What are administrative divisions called in Swaziland?
Districts
2595. How many districts does Swaziland have?
4
2596. What is the climate of Swaziland?
Varies from tropical to near temperate
2597. What are the natural resources of Swaziland?
Asbestos, coal, clay, cassiterite, hydropower, forests, small gold and diamond deposits, quarry stone, and talc
2598. What are the natural hazards of Swaziland?
Drought

2599. What are the religions of Swaziland?
Zionist 40% (a blend of Christianity and indigenous ancestral worship), Roman Catholic 20%, Muslim 10%, other (includes Anglican, Bahai, Methodist, Mormon, Jewish) 30%
2600. What are the ethnic groups of Swaziland?
African 97%, European 3%

Tanzania

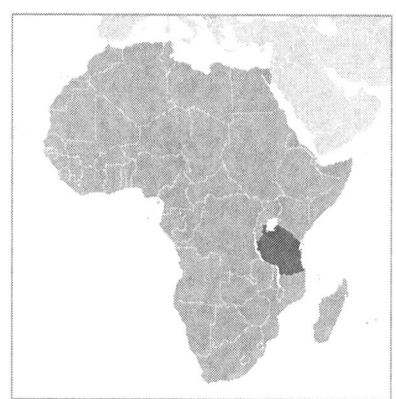

2601. What is the official name of Tanzania?
United Republic of Tanzania
2602. Which countries border Tanzania to the west?
Rwanda, Burundi and the Democratic Republic of the Congo
2603. Which countries border Tanzania to the north?
Kenya and Uganda
2604. Which countries border Tanzania to the south?
Zambia, Malawi and Mozambique
2605. Which water body borders Tanzania to the east?
Indian Ocean
2606. Tanzania is bordered by which three of the largest lakes in Africa?
Lake Victoria in the north, Lake Tanganyika in the west, and Lake Nyasa (also called Lago Niassa, Lake Malawi, and Lake Nyassa) in the southwest
2607. What is the motto of Tanzania?
Freedom and Unity
2608. What is the national anthem of Tanzania?

God Bless Africa
2609. What is the capital of Tanzania?
Dodoma
2610. What is the largest city of Tanzania?
Dar es Salaam
2611. What are official languages of Tanzania?
Kiswahili (Swahili) and English
2612. What kind of government does Tanzania have?
Republic
2613. Tanganyika was part of which colony during February 27th, 1885–June 28th, 1919?
German East Africa
2614. What was Maji Maji Rebellion (also called Maji Maji War)?
A violent African resistance to colonial rule in the German colony of Tanganyika during 1905–1907
2615. When did Tanganyika gain independence from United Kingdom-administered UN trusteeship?
December 9th, 1961
2616. When did Zanzibar gain independence from United Kingdom?
December 19th, 1963
2617. When did Tanganyika unite with Zanzibar to form United Republic of Tanganyika and Zanzibar?
April 26th, 1964
2618. When was United Republic of Tanganyika and Zanzibar renamed United Republic of Tanzania?
October 29th, 1964
2619. What kind of government is Zanzibar?
The semi-autonomous part of the United Republic of Tanzania
2620. What is the capital and largest city in Zanzibar?
Zanzibar City
2621. How large is Zanzibar?
1,020 sq mi / 2,643 km^2
2622. Zanzibar comprises which archipelago in the Indian Ocean?
Zanzibar Archipelago
2623. What is the largest island of Zanzibar Archipelago?
Unguja Island
2624. Unguja Island is surrounded by more than 20 islands, most of them uninhabited and located on the western side within which channel
Zanzibar Channel
2625. Is Mafia Island part of Zanzibar Archipelago?
No
2626. Mafia Island is consisted of how many large islands and numerous smaller ones?
1 (152 sq mi / 394 km²)
2627. What is the second largest island of Zanzibar Archipelago?
Pemba Island (380 sq mi / 984 km²)
2628. Zanzibar City is located on which island?

Unguja Island
2629. Which islands are Tanzanian Spice Islands?
Mafia Island, Unguja Island, and Pemba Island
2630. What is the area of Tanzania?
364,898 sq mi / 945,203 km²
2631. How long is the coastline of Tanzania?
885 mi / 1,424 km
2632. What is the population of Tanzania?
41,048,532 (by 2009)
2633. What is the currency of Tanzania?
Tanzanian Shilling
2634. What is the geographical feature of the terrain of Tanzania?
Plains along coast; central plateau; highlands in north, south
2635. What is the highest point of Tanzania?
The Uhuru Peak of Mount Kilimanjaro (19,341 ft / 5,895 m)
2636. What is the lowest point of Tanzania?
Indian Ocean (0 ft / 0 m)
2637. What is Mount Kilimanjaro's rank by height in Africa?
1st
2638. What is Mount Kilimanjaro's rank by height as a freestanding mountain in the world?
1st (rising 15,100 ft / 4600 m from the base)
2639. What type of mountain is Mount Kilimanjaro?
Stratovolcano
2640. Mount Kilimanjaro is composed of which three distinct volcanic cones?
Kibo (Kibo, 19,340 ft / 5895 meters), Mawenzi (extinct, 16,896 ft / 5149 m), and Shira (extinct, 13,000 ft / 3962 m)
2641. The Uhuru Peak is the highest summit on which volcanic cone's crater rim?
Kibo
2642. How large is Kilimanjaro National Park, centered on Mount Kilimanjaro?
291 square miles / 753 km²
2643. How high is Mount Meru, located 43 miles (70 km) west of Mount Kilimanjaro in Tanzania?
14,980 ft / 4,566 m
2644. What is Mount Kilimanjaro's rank by height in Africa?
5th
2645. What is the smallest national park, located in Northern Tanzania?
Gombe Stream National Park (20 sq mi / 52 km²)
2646. What is the largest national park, located in middle Tanzania?
Ruaha National Park (7799 sq mi / 20,200 km²)
2647. How large is Arusha National Park that covers Mount Meru in the north eastern Tanzania?
53 sq mi / 137 km²
2648. What type of mountain is Mount Meru?
Stratovolcano
2649. How large is Katavi National Park in western Tanzania?
1,727 sq mi / 4471 km²
2650. How large is Kitulo National Park in the Southern Highlands of Tanzania?

159 sq mi / 413 km²
2651. How large is Jozani Chwaka Bay National Park in Zanzibar?
19 sq mi / 50 km²
2652. What is the surface area of the freshwater lake, Lake Manyara, in Tanzania?
89 sq mi (231 km²)
2653. How large is Lake Manyara National Park in Tanzania?
125 sq mi / 325 km²
2654. How large is Mahale Mountains National Park, on the shores of Lake Tanganyika in western Tanzania?
637 sq mi / 1,650 km²
2655. How large is Mikumi National Park near Udzungwa Mountains and Uluguru Mountains?
1,247 sq mi / 3,230 km²
2656. How large is Mnazi Bay-Ruvuma Estuary Marine Park in the Southeast tip of Tanzania?
251 sq mi / 650 km²
2657. How many people live within Mnazi Bay-Ruvuma Estuary Marine Park?
About 30,000
2658. How large is Rubondo Island National park, located on Lake Victoria?
176 sq mi / 457 km²
2659. How large is Tarangire National Park, being crossed by the Tarangire River?
1,100 sq mi / 2,850 km²
2660. How large is Udzungwa Mountains National Park in Tanzania?
768 sq mi / 1,990 km²
2661. How large is Saadani National Park, along the Indian Ocean shores of Tanzania?
425 sq mi / 1,100 km²
2662. How large is Serengeti National Park west of Mount Kilimanjaro?
5,700 sq mi / 14,763 km²
2663. What is Serengeti National Park famous for?
Its annual migration of millions of white bearded wildebeest, as well as its abundance of lions, leopards, elephants, rhinoceroses, and buffalo
2664. What is Olduvai Gorge in the eastern Serengeti Plains in northern Tanzania?
A steep-sided ravine in the Great Rift Valley, which stretches along eastern Africa Olduvai
2665. What is Olduvai Gorge commonly referred to?
The Cradle of Mankind
2666. What are found at Olduvai Gorge?
The oldest hominid fossils and artifacts
2667. What kind of lake is Lake Natron, located in northern Tanzania?
Saline
2668. How large is Selous Game Reserve, one of the largest fauna reserves of the world, located in the south of Tanzania?
21,081 sq mi / 54,600 km²
2669. What are administrative divisions called in Tanzania?
Regions
2670. How many regions does Tanzania have?
26
2671. What is the climate of Tanzania?

Varies from tropical along coast to temperate in highlands
2672. What are the natural resources of Tanzania?
Hydropower, tin, phosphates, iron ore, coal, diamonds, gemstones, gold, natural gas, nickel
2673. What are the natural hazards of Tanzania?
Flooding on the central plateau during the rainy season; drought
2674. What are the religions of Tanzania?
Mainland - Christian 30%, Muslim 35%, indigenous beliefs 35%; Zanzibar - more than 99% Muslim
2675. What are the ethnic groups of Tanzania?
Mainland - African 99% (of which 95% are Bantu consisting of more than 130 tribes), other 1% (consisting of Asian, European, and Arab); Zanzibar - Arab, African, mixed Arab and African

Togo

2676. What is the official name of Togo?
Togolese Republic
2677. Which country borders Togo to the west?
Ghana
2678. Which country borders Togo to the north?
Burkina Faso
2679. Which country borders Togo to the east?
Benin
2680. Which water body borders Togo to the south?
Gulf of Guinea
2681. What is the motto of Togo?
Work, Liberty, Homeland
2682. What is the national anthem of Togo?
Hail to thee, land of our forefathers
2683. What is the capital of Togo?
Lomé (the largest city)
2684. What is the official language of Togo?
French
2685. What kind of government does Togo have?
Republic
2686. What was Togoland during 1884–1914?
A German protectorate including the present-day Togo and Ghana
2687. On which day was Togoland established?
July 5th, 1884
2688. What was the capital of Togoland during 1884–1886?
Bagida
2689. What was the capital of Togoland during 1886–1897?
Sebeab
2690. What was the capital of Togoland during 1897–1914?
Lomé

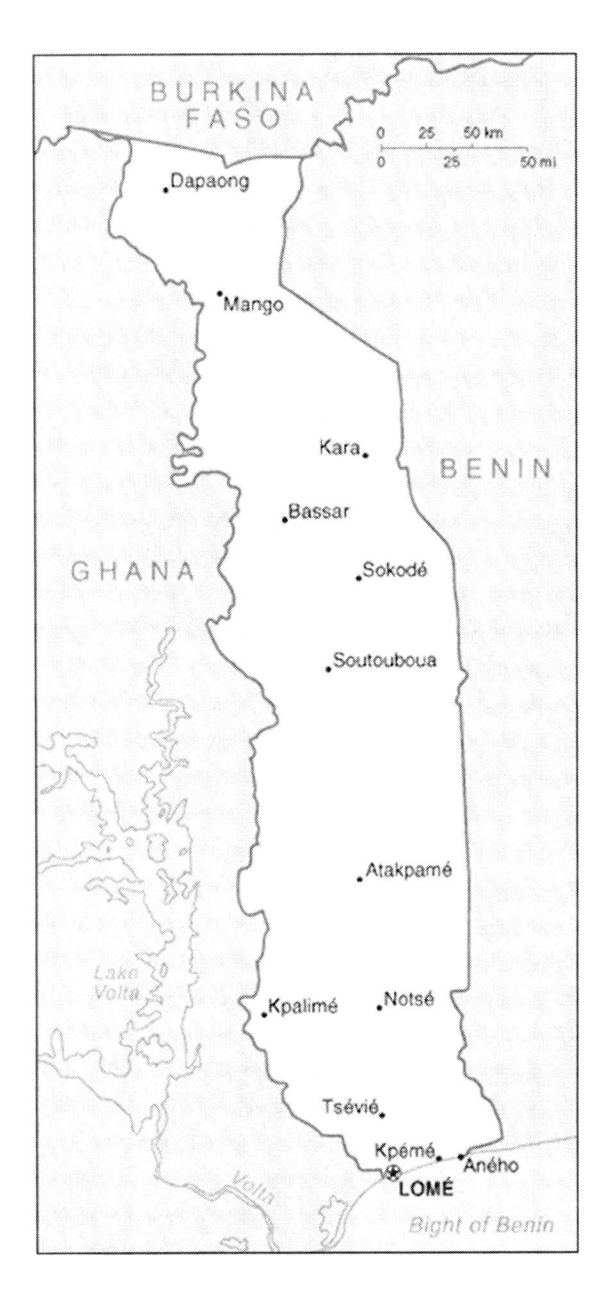

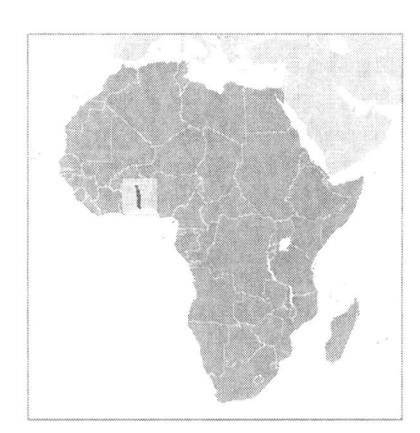

2691. What was the currency of Togoland?
German gold mark
2692. On which day was Togoland occupied by French and British troops?
August 26th, 1914
2693. Togoland was separated into which two administrative zones on December 27th, 1916?
French Togoland (the present-day Togo) and British Togoland (the present-day Ghana)
2694. With the ratification of the Treaty of Versailles in 1920, Togoland formally became what?
The League of Nations Class B mandate
2695. What was the capital of French Togoland?
Lomé

2696. After World War II, what was the legal status of French Togoland?
French-administered UN trusteeship
2697. When did Togo gain independence from French-administered UN trusteeship?
April 27th, 1960
2698. What is the area of Togo?
21,925 sq mi / 56,785 km^2
2699. How long is the coastline of Togo?
35 mi / 56 km
2700. What is the population of Togo?
6,031,808 (by 2009)
2701. What is the currency of Togo?
CFA Franc
2702. What is the geographical feature of the terrain of Togo?
Gently rolling savanna in north; central hills; southern plateau; low coastal plain with extensive lagoons and marshes
2703. What is the highest point of Togo?
Mont Agou (3,235 ft / 986 m)
2704. What is the lowest point of Togo?
Atlantic Ocean (0 ft / 0 m)
2705. What is the largest lake of Togo?
Lake Togo
2706. How long is Mono River in eastern Togo?
250 mi / 400 km
2707. The Atakora Mountains in Benin is called what in Togo?
Togo Mountains
2708. Togo contains what national parks?
Fazao-Malfakassa National Park, Fosse aux Lions National Park, and Kéran National Park
2709. What are administrative divisions called in Togo?
Regions
2710. How many districts does Togo have?
5
2711. What is the climate of Togo?
Tropical; hot, humid in south; semiarid in north
2712. What are the natural resources of Togo?
Phosphates, limestone, marble, arable land
2713. What are the natural hazards of Togo?
Hot, dry harmattan wind can reduce visibility in north during winter; periodic droughts
2714. What are the religions of Togo?
Christian 29%, Muslim 20%, indigenous beliefs 51%
2715. What are the ethnic groups of Togo?
African (37 tribes; largest and most important are Ewe, Mina, and Kabre) 99%, European and Syrian-Lebanese less than 1%

Tunisia

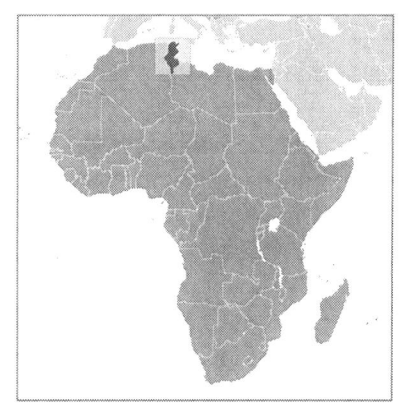

2716. What is the official name of Tunisia?
Tunisian Republic
2717. What is Tunisia's position in Africa?
Northernmost
2718. Which country borders Tunisia to the west?
Algeria
2719. Which country borders Tunisia to the southeast?
Libya
2720. Which water body borders Tunisia to the north and east?
Mediterranean Sea

2721. What is the motto of Tunisia?
Liberty, Order, Justice
2722. What is the national anthem of Tunisia?
Defenders of the Homeland
2723. What is the capital of Tunisia?
Tunis (the largest city)
2724. What is the official language of Tunisia?
Arabic
2725. What kind of government does Tunisia have?
Republic
2726. What is Carthage?
Carthage refers to a series of cities on the Gulf of Tunis, from a Phoenician colony of the 1st millennium BCE to the current suburb outside Tunis, Tunisia
2727. What happened to Carthage after the First Punic War during 264 BCE–241 BCE with Rome?
Carthage signed a peace treaty under the terms of which they evacuated Sicily and paid Rome a large war indemnity
2728. What happened to Carthage after the Second Punic War (also called the Hannibalic War, or the War Against Hanniba) during 218 BCE–201 BCE with Rome?
Carthage lost Hispania forever, and it was reduced to a Roman client state
2729. Berbers was on which side during the Second Punic War?
Carthage's
2730. Which kingdom, in present-day Algeria and part of Tunisia, was formed by Berbers during 202 BCE–46 BCE?
Numidia
2731. What happened to Carthage after the Third Punic War during 149 BCE–146 BCE with Rome?
Carthage become the Roman province of Africa
2732. Numidia was on which side during the Third Punic War?
Rome's
2733. Carthage and its Africa province were captured who in 439?
Vandals
2734. Who recaptured Rome's Africa province during the Vandalic War in 534?
The Eastern Romans (also called Byzantine Empire)
2735. Who entered the region of Roman Province of Africa in 670?
Arab Muslim
2736. Roman Province of Africa was called what in Arabic?
Ifriqiya
2737. Who ruled Ifriqiya in 1574?
Ottoman Empire
2738. When was Tunisia officially made a French protectorate with the Treaty of Bardo?
May 12th, 1881
2739. When did Tunisia gain independence from France?
March 20, 1956
2740. What is the area of Tunisia?
63,170 sq mi / 163,610 km^2

2741. How long is the coastline of Tunisia?
   713 mi / 1,148 km
2742. What is the population of Tunisia?
   10,486,339 (by 2009)
2743. What is the currency of Tunisia?
   Tunisian Dinar
2744. What is the geographical feature of the terrain of Tunisia?
   Mountains in north; hot, dry central plain; semiarid south merges into the Sahara
2745. What is the highest point of Tunisia?
   Jebel ech Chambi (5,066 ft / 1,544 m)
2746. What is the lowest point of Tunisia?
   Shatt al Gharsah (-56 ft / -17 m)
2747. Which gulf, located on Tunisia's east coast in the Mediterranean Sea, was called Minor Syrtis in ancient times?
   Gulf of Gabès
2748. Which group of islands was on the northeast of Gulf of Gabes?
   Kerkennah Islands
2749. Which island was on the southeast of Gulf of Gabes?
   Djerba
2750. What was Djerba's rank by size in North Africa?
   1st (198 sq mi / 514 km²)
2751. Tunisia contains what national parks?
   Bou-Hedma National Park, Boukornine National Park, Chaambi National Park, Feija National Park, Ichkeul National Park, Sidi Toui National Park, Zembra National Park, and Jebil National Park
2752. What are administrative divisions called in Tunisia?
   Governorates
2753. How many governorates does Tunisia have?
   29
2754. What is the climate of Tunisia?
   Temperate in north with mild, rainy winters and hot, dry summers; desert in south
2755. What are the natural resources of Tunisia?
   Petroleum, phosphates, iron ore, lead, zinc, salt
2756. What are the religions of Tunisia?
   Muslim 98%, Christian 1%, Jewish and other 1%
2757. What are the ethnic groups of Tunisia?
   Arab 98%, European 1%, Jewish and other 1%

Uganda

2758. What is the official name of Uganda?
   Republic of Uganda
2759. Which country borders Uganda to the southwest?
   Rwanda

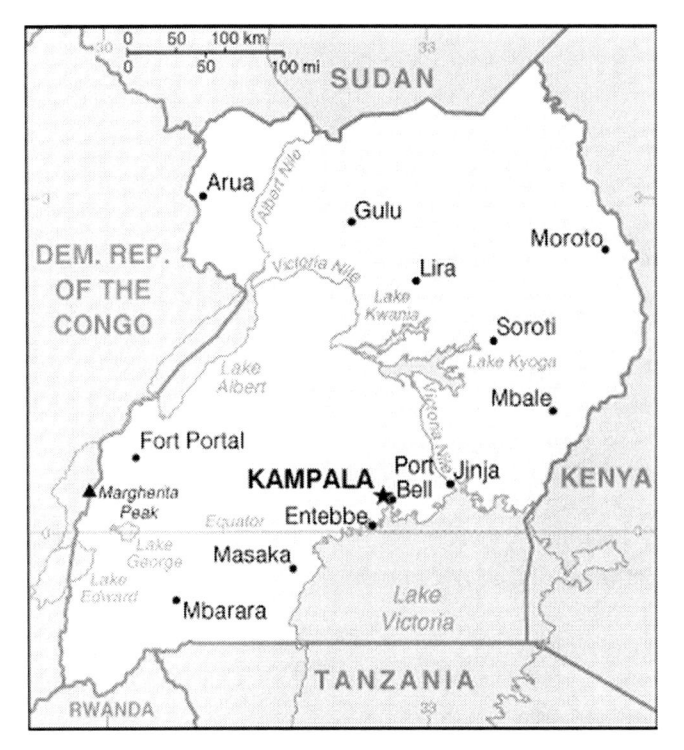

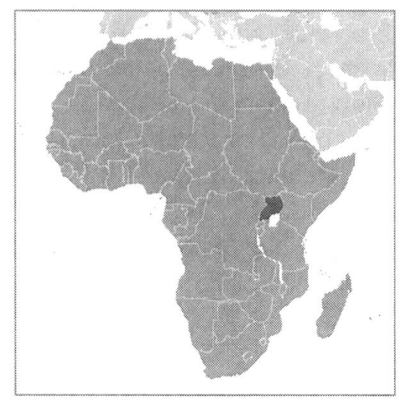

2760. Which country borders Uganda to the south?
Tanzania
2761. Which country borders Uganda to the east?
Kenya
2762. Which country borders Uganda to the north?
Sudan
2763. Which country borders Uganda to the west?
Democratic Republic of the Congo
2764. Which lake is on the southern border?
Lake Victoria
2765. What is the motto of Uganda?
For God and My Country
2766. What is the national anthem of Uganda?
Oh Uganda, Land of Beauty
2767. What is the capital of Uganda?
Kampala (the largest city)
2768. What is the official language of Uganda?
English, Swahili
2769. What kind of government does Uganda have?
Democratic Republic
2770. What are five constituent kingdoms of Uganda?
Buganda, Bunyoro-Kitara, Ankole, Toro, and Busoga
2771. What is the largest constituent kingdom of Uganda?
Buganda Kingdom

2772. Buganda is the Swahili term for what?
Uganda
2773. Kampala is located in which constituent kingdom of Uganda?
Buganda Kingdom
2774. Where is Buganda Kingdom?
It is bounded on the south by Lake Victoria, on the east by White Nile, on the north by Lake Kyoga, on the northwest by River Kafu, and on the west by the districts of Isingiro, Kiruhura, Kyenjojo, Kibale, Hoima and Masindi
2775. Buganda Kingdom is compromised by which ethnic people?
Baganda
2776. When did Baganda become Uganda Protectorate of United Kingdom?
1894
2777. Is it a fact that Kitara Empire was a kingdom which, at the height of its power in 14th and 15th centuries, included much of the present-day Uganda, northern Tanzania and eastern Democratic Republic of the Congo?
It is in the oral tradition
2778. Which kingdom was created when the ancient Empire of Kitara broke apart during 16th century?
Bunyoro-Kitara
2779. Bunyoro-Kitara Kingdom controlled almost the entire region between Lake Victoria, Lake Edward, and Lake Albert until which century?
19th
2780. In around 1830, which large province was separated by a prince of Bunyoro from Bunyoro-Kitara Kingdom to become a separate kingdom?
Toro
2781. In July 1890 agreement, Bunyoro-Kitara Kingdom gave the entire region north of Lake Victoria to which country and later the region was declared to be its protectorate in 1894?
United Kingdom
2782. In which year, Toro Kingdom was incorporated back into Bunyoro-Kitara Kingdom?
1876
2783. When did Toro Kingdom reassert its independence?
1891
2784. Which kingdom, established in late 15th century, is located in the southwestern Uganda, east of Lake Edward?
Ankole (also called Nkore)
2785. When was Ankole Kingdom incorporated into the British protectorate of Uganda by the signing of the Ankole agreement?
October 25th, 1901
2786. Which kingdom, technically one of the constituent kingdoms, is not a formal monarchy?
Busoga
2787. Why was Busoga Kingdom established?
It is established as a cultural institution that promotes popular participation and unity among the people of area
2788. Where is Busoga Kingdom?

It is bounded on the north by the swampy Lake Kyoga which, on the west by the White Nile, on the south by Lake Victoria, and on the east by the Mpologoma (also called Lion) River. It also includes some islands in Lake Victoria

2789. Who was the first ruler of Busoga Kingdom?
Ezekiel Tenywa Wako (February 11, 1939–1949)

2790. Traditional kingdoms were abolished in 1967 by whom?
President, Milton Obote

2791. When were constituent kingdoms restored?
1993

2792. Which kingdom did not have an official undisputed king because of disagreements between the ruling clans?
Ankole Kingdom

2793. When did Uganda gain independence from United Kingdom?
October 9th, 1962

2794. The Uganda-Tanzania War (also called the Liberation War) was fought between Uganda and Tanzania in during which period?
1978–1979

2795. Which country was at Uganda's side?
Libya

2796. What was the result of the Uganda-Tanzania War?
Tanzanian victory; overthrow of the president, Idi Amin, in Uganda

2797. What was the Ugandan Bush War (also called the war in the bush, and or the Luwero War or the Ugandan civil war)?
It refers to the guerrilla war waged during 1981–1986 by the National Resistance Army (NRA) against then presidents

2798. What was the result of the Ugandan Bush War?
Victory for National Resistance Army: Leader, Museveni, became president; NRA became the national army and was renamed Uganda People's Defence Force (UPDF)

2799. What is Luwero triangle in Uganda?
An area, to the north of the Kampala, where Ugandan Bush War started

2800. What is Lord's Resistance Army (also called Lord's Resistance Movement or Lakwena Part Two)?
A sectarian Christian militant group based in northern Uganda

2801. The LRA is currently proscribed as what kind organization by United States?
The terrorist organization

2802. What was the Lord's Resistance Army insurgency?
A guerrilla campaign waged during 1987–2009 by the LRA, operating mainly in northern Uganda, but also in South Sudan and eastern DR Congo

2803. What was the status of the Lord's Resistance Army insurgency?
The LRA was militarily defeated. All military bases in South Sudan and Northern Uganda were destroyed. Some fighters might continue to commit attacks on Democratic Republic of the Congo civilians in the East

2804. What is the area of Uganda?
91,136 sq mi / 236,040 km^2

2805. How long is the coastline of Uganda?

0 mi / 0 km
2806. What is the population of Uganda?
32,369,558 (by 2009)
2807. What is the currency of Uganda?
Ugandan shilling
2808. What is the geographical feature of the terrain of Uganda?
Mostly plateau with rim of mountains
2809. What is the highest point of Uganda?
Margherita Peak on Mount Stanley (16,765 ft / 5,110 m)
2810. Mount Stanley is also the highest mountain of which other country?
Democratic Republic of the Congo
2811. What is the lowest point of Uganda?
Lake Albert (2,037 ft / 621 m)
2812. Which lake is a large shallow lake complex in the middle of Uganda?
Lake Kyoga
2813. What is the surface area of Lake Kyoga?
664 sq mi / 1,720 km²
2814. Which river flows through Lake Kyoga on its way from Lake Victoria to Lake Albert?
White Nile
2815. What is the outflow of Lake Kyoga?
Murchison Falls
2816. What is the biggest national park in Uganda?
Murchison Falls National Park (1,340 sq mi / 3,480 km²)
2817. What is Uganda's only National Park which has all "big five" (lion, leopard, buffalo, rhinoceros and elephant)?
Murchison Falls National Park
2818. What is Uganda's most-visited game reserve?
Queen Elizabeth National Park
2819. How large is Queen Elizabeth National Park, home to 95 species of mammal and over 500 species of birds, located in western Uganda?
764 sq mi / 1,978 km²
2820. How large is Bwindi Impenetrable National Park, , located in southwestern Uganda, which is a habitat for some 120 species of mammals, 346 species of birds, 202 species of butterflies, 163 species of trees, 100 species of ferns, 27 species of frogs, chameleons, geckos and many endangered species?
128 sq mi / 331 km²
2821. How large is Katonga Wildlife Reserve, located in western Uganda, which is named after Katonga River?
81 sq mi /211 km²
2822. What wildlife species exist in Katonga Wildlife Reserve?
Sitatunga, Reedbuck, Waterbuck, Warthog, Bushbuck, Colobus monkey, Elephant, and River otter
2823. How large is Kibale National Park, protecting moist evergreen rain forest?
300 sq mi / 776 km²

2824. How large is Kidepo Valley National Park, located in northeast Uganda, which is dominated by Mount Morungole (9,020 ft / 2,750 m) and transected by River Kidepo and River Narus?
557 sq mi / 1,442 km^2

2825. How large is Lake Mburo National Park, located in western Uganda, which is home to variety of animals such as zebras, impala, and buffaloes?
100 sq mi / 260 km^2

2826. How large is Mgahinga Gorilla National Park in the far southwestern Uganda?
13 sq mi / 34 km^2

2827. Mgahinga Gorilla National Park and which other two national parks form the Virunga Conservation Area?
Virunga National Park (Democratic Republic of the Congo) and Volcanoes National Park (Rwanda)

2828. Mount Elgon National Park is shared by Uganda and which other country?
Kenya

2829. How large is Rwenzori Mountains National Park, located in southwestern Uganda, which is known for glaciers, snowfields, waterfalls, and lakes?
385 sq mi / 998 km^2

2830. How large is Semuliki National Park, located in western Uganda, which is East Africa's only lowland tropical rainforest?
85 sq mi / 220 km^2

2831. Which two water falls on White Nile near Lake Victoria were submerged in 1954 with the completion of the Owen Falls Dam?
Owen Falls and Ripon Falls

2832. What are administrative divisions called in Uganda?
Districts

2833. How many districts does Uganda have?
80

2834. What is the climate of Uganda?
Tropical; generally rainy with two dry seasons (December to February, June to August); semiarid in northeast

2835. What are the natural resources of Uganda?
Copper, cobalt, hydropower, limestone, salt, arable land, gold

2836. What are the religions of Uganda?
Roman Catholic 41.9%, Protestant 42% (Anglican 35.9%, Pentecostal 4.6%, Seventh Day Adventist 1.5%), Muslim 12.1%, other 3.1%, none 0.9%

2837. What are the ethnic groups of Uganda?
Baganda 16.9%, Banyakole 9.5%, Basoga 8.4%, Bakiga 6.9%, Iteso 6.4%, Langi 6.1%, Acholi 4.7%, Bagisu 4.6%, Lugbara 4.2%, Bunyoro 2.7%, other 29.6%

Western Sahara

2838. Which country borders Western Sahara to the northeast?
Algeria

2839. Which country borders Western Sahara to the north?
Morocco

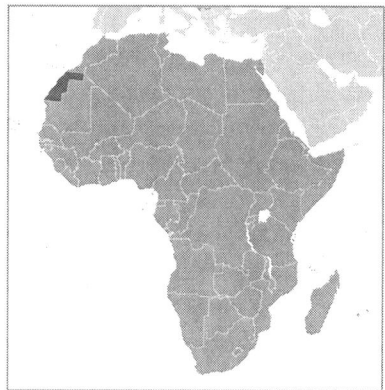

2840. Which country borders Western Sahara to the east and south?
   Mauritania
2841. Which water body borders Western Sahara to the west?
   Atlantic Ocean
2842. What is the largest city of Western Sahara?
   El Aaiún
2843. What is the official language of Western Sahara?
   Arabic
2844. What was Spanish Sahara?
   The territory of Western Sahara when it was ruled as a territory by Spain
2845. When was Spanish Sahara established?
   December 26th, 1884
2846. What was the capital of Spanish Sahara?
   El Aaiún
2847. What was the currency of Spanish Sahara?
   Spanish Peseta
2848. What was the Green March?
   A strategic mass demonstration in November 1975, coordinated by the Moroccan government, to force Spain to hand over the control of Western Sahara to Morocco
2849. What was the Madrid Accords (also called Madrid Agreement or Madrid Pact)?
   A treaty signed in Madrid between Spain, Morocco, and Mauritania to end the Spanish presence in the territory of Spanish Sahara
2850. When was Madrid Accords signed?
   November 14th, 1975

2851. Which country annexed the southern third of Spanish Sahara, and named it Tiris al-Gharbiyya in 1975 after the Madrid Accords?
Mauritania

2852. Which country annexed the northern two-thirds of Spanish Sahara as its Southern Provinces in 1976?
Morocco

2853. When was Spanish Sahara disestablished?
February 26th, 1976

2854. What organization is Polisario Front?
A Sahrawi rebel national liberation movement working for the independence of Western Sahara

2855. Polisario Front proclaimed which country on February 27, 1976?
Sahrawi Arab Democratic Republic

2856. When did Polisario Front force Mauritania to withdraw from Western Sahara?
1979

2857. How did the guerrilla war between Polisario Front and Morocco end in 1991?
By UN-brokered cease-fire

2858. What is the current status of Western Sahara?
The UN-organized referendum on the territory's final status has been repeatedly postponed

2859. What does Polisario Front want?
The popular referendum that includes the option of independent Western Sahara

2860. What has Morocco proposed?
The autonomy proposal for the territory, which would allow for some local administration while maintaining Moroccan sovereignty

2861. What is Berm of Western Sahara (also called Moroccan Wall or Wall of Shame)?
It is an approximately 2,700 km-long defensive structure, mostly a sand wall, running through Western Sahara and the southeastern portion of Morocco. It acts as a separation barrier between Polisario Front and Morocco.

2862. What is the area of Western Sahara?
102,703 sq mi / 266,000 km^2

2863. How long is the coastline of Western Sahara?
690 mi / 1,110 km

2864. What is the population of Western Sahara?
405,210 (by 2009)

2865. What are currencies of Western Sahara?
Moroccan Dirham, Sahrawi Peseta

2866. What is the geographical feature of the terrain of Western Sahara?
Mostly low, flat desert with large areas of rocky or sandy surfaces rising to small mountains in south and northeast

2867. What is the highest point of Western Sahara?
Unnamed elevation (2,641 ft / 805 m)

2868. What is the lowest point of Western Sahara?
Sebjet Tah (180 ft / -55 m)

2869. What are administrative divisions called in Western Sahara?

None
2870. What is the climate of Western Sahara?
Hot, dry desert; rain is rare; cold offshore air currents produce fog and heavy dew
2871. What are the natural resources of Western Sahara?
Phosphates, iron ore
2872. What are the natural hazards of Western Sahara?
Hot, dry, dust/sand-laden sirocco wind can occur during winter and spring; widespread harmattan haze exists 60% of time, often severely restricting visibility
2873. What are the religions of Western Sahara?
Muslim
2874. What are the ethnic groups of Western Sahara?
Arab, Berber

Zambia

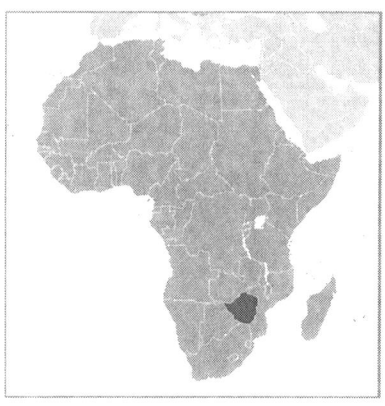

2875. What is the official name of Zambia?
Republic of Zambia
2876. Which country borders Zambia to the northwest?
Democratic Republic of the Congo
2877. Which country borders Zambia to the west?
Angola
2878. Which country borders Zambia to the northeast?
Tanzania
2879. Which country borders Zambia to the east?

Malawi
2880. Which countries border Zambia to the south?
Mozambique, Zimbabwe, Botswana, and Namibia
2881. Which river forms the border between Zambia and Zimbabwe?
Zambezi River
2882. What is the motto of Zambia?
One Zambia, One Nation
2883. What is the national anthem of Zambia?
Stand and Sing of Zambia, Proud and Free
2884. What is the capital of Zambia?
Lusaka (the largest city)
2885. What are official languages of Zambia?
Bemba, Nyanja, Tonga, Lozi, Lunda, Kaonde, Luvale, and English
2886. What kind of government does Zambia have?
Republic
2887. The territory of Northern Rhodesia was administered by which organization from 1891 until it was taken over by United Kingdom on April 1, 1924
British South Africa Company
2888. What was the capital of Northern Rhodesia until 1935?
Livingstone
2889. What was the capital of Northern Rhodesia from 1935?
Lusaka
2890. During which period Northern Rhodesia was part of Federation of Rhodesia and Nyasaland?
August 1st, 1953–December 31st, 1963
2891. When did Northern Rhodesia gain independence from United Kingdom with the new name of Zambia?
October 24th, 1964
2892. Zambia was governed as single-party rule by whom for 27 years?
President Kenneth Kaunda
2893. How did Zambia become a multiparty democracy from 1991?
Peaceful multiparty elections
2894. What is the area of Zambia?
290,587 sq mi / 752,618 km^2
2895. How long is the coastline of Zambia?
0 mi / 0 km
2896. What is the population of Zambia?
11,862,740 (by 2009)
2897. What is the currency of Zambia?
Zambian kwacha
2898. What is the geographical feature of the terrain of Zambia?
Mostly high plateau with some hills and mountains
2899. What is the highest point of Zambia?
Unnamed location in Mafinga Hills (7,549 ft / 2,301 m)
2900. What is the lowest point of Zambia?
Zambezi River (1,079 ft / 329 m)

2901. Which two of the Zambezi's longest and largest tributaries flow mainly in Zambia?
Kafue River and Luangwa River
2902. What is Barotse Floodplain (also called Bulozi Plain, Lyondo or Zambezi Floodplain)?
It is one of Africa's great wetlands, on the Zambezi River in the Western Province of Zambia
2903. What is the largest national park in Zambia, which is named after a river?
Kafue National Park (8,649 sq mi / 22,400 km²)
2904. How large is Blue Lagoon National Park, located in Central Province of Zambia?
174 sq mi / 450 km²
2905. How large is Kasanka National Park, located in Central Province of Zambia, which is managed and owned by the Kasanka Trust?
174 sq mi / 450 km²
2906. How large is Liuwa Plain National Park, located in Western Province of Zambia, which is designated as a game reserve?
1,390 sq mi / 3,600 km² (including vast wooded islands, and the Liuwa Plain)
2907. How large is Lochinvar National Park, located south west of Lusaka, which is on the south side of the Kafue River?
165 sq mi / 428 km²
2908. How large is Lower Zambezi National Park, located on the north bank of the Zambezi River in south eastern Zambia?
1,579 sq mi / 4,090 km²
2909. How large is Luambe National Park, located in the Eastern Province of Zambia?
98 sq mi / 254 km²
2910. How large is Lukusuzi National Park, located in eastern Luangwa Valley in Zambia?
1,050 sq mi / 2,720 km²
2911. How large is Mosi-oa-Tunya National Park, located on the Zambezi River, which is known for Victoria Falls?
25 sq mi / 66 km²
2912. How large is Mweru Wantipa National Park, located in the Northern Province of Zambia, which is named after Lake Mweru Wantipa?
1,210 sq mi / 3,134 km²
2913. How large is North Luangwa National Park, located in the valley of the Luangwa River?
1,790 sq mi / 4,636 km²
2914. How large is Sumbu National Park (also called Nsumbu), located in Northern Province of Zambia, which is on the western shore of Lake Tanganyika?
780 sq mi / 2,020 km²
2915. How large is Nyika National Park, located in the northeast of Zambia, which has the approximate altitude of 6,562 ft / 2,000 m?
31 sq mi / 80 km²
2916. How large is Sioma Ngwezi National Park, located in the southwestern corner of Zambia, which is along the border of Namibia and Angola?
2,037 sq mi / 5276 km²
2917. How large is South Luangwa National Park, located in the valley of the Luangwa River in eastern Zambia, which is a game reserve?
3,494 sq mi / 9,050 km²

2918. South Luangwa National Park has over 400 bird species and 60 different animal species. Among them what is unique to Luangwa valley?
Thornicroft's Giraffe
2919. How large is West Lunga National Park, which is a remote wildlife haven in dense forest in the Northwestern Province of Zambia?
650 sq mi / 1684 km²
2920. What are administrative divisions called in Zambia?
Provinces
2921. How many provinces does Zambia have?
9
2922. What is the climate of Zambia?
Tropical; modified by altitude; rainy season (October to April)
2923. What are the natural resources of Zambia?
Copper, cobalt, zinc, lead, coal, emeralds, gold, silver, uranium, hydropower
2924. What are the natural hazards of Zambia?
Periodic drought; tropical storms (November to April)
2925. What are the religions of Zambia?
Christian 50%-75%, Muslim and Hindu 24%-49%, indigenous beliefs 1%
2926. What are the ethnic groups of Zambia?
African 99.5% (includes Bemba, Tonga, Chewa, Lozi, Nsenga, Tumbuka, Ngoni, Lala, Kaonde, Lunda, and other African groups), other 0.5% (includes Europeans, Asians, and Americans)

Zimbabwe

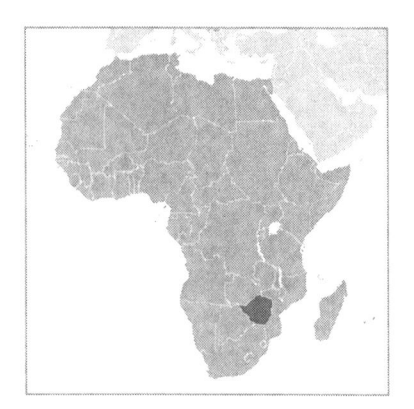

2927. What is the official name of Zimbabwe?
Republic of Zimbabwe
2928. Which country borders Zimbabwe to the southwest?
Botswana
2929. Which country borders Zimbabwe to the northwest?
Zambia
2930. Which country borders Zimbabwe to the south?
South Africa
2931. Which country borders Zimbabwe to the east?
Mozambique
2932. What is the motto of Zimbabwe?
Unity, Freedom, Work
2933. What is the national anthem of Zimbabwe?
Blessed be the land of Zimbabwe
2934. What is the capital of Zimbabwe?
Harare (the largest city)
2935. What is the official language of Zimbabwe?
English
2936. What kind of government does Zimbabwe have?
Semi presidential, parliamentary, consociationalist republic
2937. The territory of Northern Rhodesia was administered by which organization from 1891 until it was taken over by United Kingdom on April 1, 1924
British South Africa Company
2938. Queen Victoria signed the charter in 1889 for whom to administer Southern Rhodesia as a protectorate?
British South Africa Company
2939. On which day did Southern Rhodesia become a self-governing colony?
October 1st, 1923
2940. What was the capital of Southern Rhodesia?
Salisbury
2941. What was the currency of Southern Rhodesia?
Southern Rhodesian Pound
2942. During which period Southern Rhodesia was part of Federation of Rhodesia and Nyasaland?
August 1st, 1953–December 31st, 1963
2943. When was Unilateral Declaration of Independence (UDI) of Rhodesia from the United Kingdom signed?
November 11th, 1965
2944. Was Republic of Rhodesia recognized?
No, United Kingdom, Commonwealth, and United Nations condemned the move as illegal
2945. What was the motto of Rhodesia?
May she be worthy of the name
2946. What was the national anthem of Rhodesia from 1974?
Rise O Voices of Rhodesia
2947. What was the capital of Rhodesia?
Salisbury

2948. What was the currency of Rhodesia until 1970?
Rhodesian Pound
2949. What was the currency of Rhodesia from 1970?
Rhodesian Dollar
2950. What was Rhodesian Bush War (also called Zimbabwe War of Liberation or Second Chimurenga)?
A civil war in Rhodesia fought from July 1964 to 1979
2951. On December 1st, 1979, delegations from the British and Rhodesian governments and the Patriotic Front signed which agreement to end Rhodesian Bush War
Lancaster House Agreement
2952. What is Zimbabwe Rhodesia?
An unrecognized state that existed from June 1st, 1979 to December 12th, 1979
2953. What was the capital of Zimbabwe Rhodesia?
Salisbury
2954. What was the currency of Zimbabwe Rhodesia?
Rhodesian Dollar
2955. Rhodesia reverted to de facto British control as the British Dependency of Southern Rhodesia during which period?
December 13th, 1979–April 17th, 1980
2956. When did Southern Rhodesia gain independence from United Kingdom with the new name of Zimbabwe?
April 18th, 1980
2957. What is the area of Zimbabwe?
150,871 sq mi / 390,757 km^2
2958. How long is the coastline of Zimbabwe?
0 mi / 0 km
2959. What is the population of Zimbabwe?
11,392,629 (by 2009)
2960. What is the currency of Zimbabwe?
US Dollar
2961. What is the geographical feature of the terrain of Zimbabwe?
Mostly high plateau with higher central plateau (high veld); mountains in east
2962. What is the largest lake in Zimbabwe?
Lake Kariba (2,154 sq mi / 5,580 km^2)
2963. By which measure, Lake Kariba is the largest man-made lake and reservoir in the world?
Volume (43 cube mi / 180 km^3)
2964. Which dam completed in 1959 made Lake Kariba between Zambia and Zimbabwe?
Kariba Dam
2965. What is the highest point of Zimbabwe?
Nyangani (also called Inyangani, 8,504 ft / 2,592 m)
2966. What is the lowest point of Zimbabwe?
Junction of Runde River and Save River (531 ft / 162 m)
2967. Mount Nyangani is located in which national park?
Nyanga National Park (182 sq mi / 472 km²)

2968. Which water fall is located in Nyanga National Park in the Eastern Highlands and is Zimbabwe's highest waterfall?
Mutarazi Falls (2,499 ft / 762 m)
2969. What is the largest game reserve in Zimbabwe, and one of Africa's elephant strongholds?
Hwange National Park (also called Wankie, 5,657 sq mi / 14,651 km²)
2970. How large is Chimanimani National Park, located in southeastern Zimbabwe, which is famous for spectacular scenery with waterfalls, streams and virgin forests?
66 sq mi / 171 km²
2971. What is Chinhoyi Caves National Park in northern Zimbabwe famous for?
The cave system composed of limestone and dolomite, and the descent to the main cave with its pool of cobalt blue water is very impressive
2972. How large is Gonarezhou National Park located in south eastern Zimbabwe?
1,951 sq mi / 5,053 km²
2973. Gonarezhou National Park, Limpopo National Park in Mozambique and Kruger *National* Park in South Africa are part of which peace park?
Great Limpopo Transfrontier Park (13,514 sq mi / 35,000 km²)
2974. How large is Kazuma Pan National Park, located in the extreme northwest corner of Zimbabwe, which is a haven for birdlife, as well as home to roan antelope, tsessebe, cheetah, rhinoceros, giraffe and many other species?
121 sq mi / 313 km²
2975. How large is Mana Pools National Park, located in northern Zimbabwe, which is famous for large population of crocodiles, hippopotamuses and elephants?
846 sq mi / 2,190 km²
2976. How large is Matobo National Park, located in southern Zimbabwe, which is famous for the high concentration of black eagles and rhinoceros, hosting the crown eagle?
172 sq mi / 445 km²
2977. Matobo National Park forms the core of which mountain range, which has the most majestic granite scenery, domes and remarkable rock formations?
Matobo Hills (also called Matopos Hills)
2978. How large is Matusadona National Park, located in western Zimbabwe, which is a game reserve park that is famous for high concentrations of lions and many other mammals?
543 sq mi / 1,407 km²
2979. Matusadona National Park is on southern shores of which lake?
Lake Kariba
2980. How large is Thuli Parks and Wildlife Land, a protected area in southwestern Zimbabwe?
161 sq mi / 416 km²
2981. The Thuli Parks and Wildlife Land comprises the Thuli Safari Area, plus which three small botanical reserves?
Pioneer Botanical Reserve, South Camp Botanical Reserve, and Tolo River Botanical Reserve
2982. How large is Victoria Falls National Park in northwestern Zimbabwe?
9 sq mi / 23 km²
2983. How is Victoria Falls claimed to be the largest in the world?
It is based on a width of 5,604 ft (1,708 m) and height of 360 ft (108 m), forming the largest sheet of falling water

2984. How large is Zambezi National Park, located in upstream Victoria Falls, which stretches along Zambezi River over 40 km?
217 sq mi / 562 km²
2985. What is Zambezi National Park famous for?
The "big five" (lion, leopard, buffalo, rhinoceros and elephant) and a wide variety of other animals, such as, giraffes and sable antelopes
2986. What are administrative divisions called in Zimbabwe?
Provinces
2987. How many provinces does Zimbabwe have?
8 (plus 2 cities)
2988. What is the climate of Zimbabwe?
Tropical; moderated by altitude; rainy season (November to March)
2989. What are the natural resources of Zimbabwe?
Coal, chromium ore, asbestos, gold, nickel, copper, iron ore, vanadium, lithium, tin, platinum group metals
2990. What are the natural hazards of Zimbabwe?
Recurring droughts; floods and severe storms are rare
2991. What are the religions of Zimbabwe?
Syncretic (part Christian, part indigenous beliefs) 50%, Christian 25%, indigenous beliefs 24%, Muslim and other 1%
2992. What are the ethnic groups of Zimbabwe?
African 98% (Shona 82%, Ndebele 14%, other 2%), mixed and Asian 1%, white less than 1%

Miscellaneous

2993. What are banks and shoals?
Once volcanic islands which have now sunk, or eroded to below sea level, or to low coral islands
2994. What is the world's largest desert?
The Sahara Desert
2995. How large is the Sahara Desert?
3,500,000 sq mi / 9,064,958 km²
2996. What are Sahara's boundaries?
Atlantic Ocean on the west, Atlas Mountains and the Mediterranean Sea on the north, Red Sea and Egypt on the east, and Sudan and the valley of the Niger River on the south
2997. What is the highest peak in the Sahara?
Emi Koussi (11,200 ft / 3,415 m)
2998. The Sahara divides the Africa into which two parts?
North Africa and Sub-Saharan Africa
2999. The Sahara covers which countries?
Algeria, Chad, Egypt, Libya, Mali, Mauritania, Morocco, Niger, Western Sahara, Sudan and Tunisia
3000. What is Sub-Saharan Africa?
The area of the African continent which lies south of the Sahara or those African countries which are fully or partially located south of the Sahara

3001. Why is Sub-Saharan Africa called Black Africa?
Because it has many black populations
3002. Which countries or regions in Africa are not Sub-Saharan countries?
Algeria, Egypt, Libya, Morocco, Tunisia, and Western Sahara
3003. What is Horn of Africa (also called Northeast Africa, Somali Peninsula)?
A peninsula in East Africa that juts for hundreds of miles into the Arabian Sea, and lies along the southern side of the Gulf of Aden. It is the easternmost projection of the African continent.
3004. Which countries are located in the Horn of Africa?
Eritrea, Djibouti, Ethiopia, and Somalia
3005. What is Maghreb?
A region refers collectively to the countries of Morocco, Algeria, and Tunisia
3006. Which states are included in the Arab Maghreb Union?
Algeria, Libya, Mauritania, Morocco, and Tunisia
3007. What is the Sahel?
A geographical and climatic region, stretching across the north of the continent between the Atlantic Ocean and the Red Sea
3008. What is the geographic feature of the Sahel?
A transitional ecoregion of semi-arid grasslands, savannas, steppes, and thorn shrublands lying between the wooded Sudanian savanna to the south and the Sahara to the north
3009. The Sahel is located south of the Sahara desert and north of Sudan. What is the largest width of the belt?
620 mi / 998 km
3010. The Sahel covers parts of which countries?
Senegal, Mauritania, Mali, Burkina Faso, Niger, Nigeria, Chad, Sudan, Somalia, Ethiopia, and Eritrea
3011. How large is the Sahel?
1,178,800 sq mi / 3,053,200 km^2
3012. What are the Sahelian kingdoms?
A series of kingdoms or empires that were centered on the Sahel
3013. What are some Sahelian kingdoms?
Mali Empire, Songhai Empire, Kanem Empire, Ghana Empire, Wolof Empire, Bornu Empire, Oyo Empire, Benin Empire, Kaabu Empire, Lunda Empire, Bamana Empire, Wassoulou Empire
3014. The Kanem Empire was located in which present-day countries?
Chad, Libya and Niger
3015. When did Kanem Empire exist?
700–1376
3016. What was the capital of Kanem Empire?
Njimi
3017. How large was Kanem Empire?
300,000 sq mi / 776,996 km^2
3018. What language was used in Kanem Empire?
Teda
3019. What was the religion of Kanem Empire?

Islam
3020. What is Afar Depression (also called Danakil Depression or Afar Triangle)?
A geological depression near the Horn of Africa, where it overlaps Eritrea, the Afar Region of Ethiopia, and Djibouti
3021. Which desert is located in Afar Depression?
Danakil Desert
3022. Lake Assal is located in which depression?
Afar Depression
3023. The Atlas Mountains are a mountain range across a northern stretch of Africa extending about 1,500 miles (2,400 km) through which countries?
Morocco, Algeria, and Tunisia
3024. What is the highest point of the Atlas Mountains?
Jbel Toubkal (13,671 ft / 4,167 m)
3025. The Atlas Mountains are divided into which mountain ranges?
Middle Atlas, High Atlas, Anti-Atlas, Tell Atlas, and Saharan Atlas
3026. Which Atlas are only located in Morocco?
Middle Atlas, High Atlas, and Anti-Atlas
3027. Which Atlas is the lowest?
Tell Atlas
3028. Which Atlas is the largest?
Saharan Atlas
3029. Which Atlas marks the northern edge of the Sahara Desert?
Saharan Atlas
3030. What is the longest river in the world?
The Nile River
3031. How long is the Nile River?
4,132 mi / 6,650 km
3032. Which countries does the Nile River flow through?
Ethiopia, Sudan, Egypt, Rwanda, Tanzania, Uganda, Burundi, Democratic Republic of the Congo, Eritrea, and Kenya
3033. Where is the mouth if the Nile?
Mediterranean Sea
3034. The Nile has which two major tributaries?
White Nile and Blue Nile
3035. Where does White Nile and Blue Nile meet?
The Sudanese capital Khartoum
3036. White Nile rises in the Great Lakes region of central Africa. What is the elevation of the White Nile?
8,858 ft / 2,700 m
3037. The Great Lakes of Africa are a series of lakes in and around which area?
Great Rift Valley
3038. Which lakes from big to small are include in the list of the Great Lake?
Lake Victoria, Lake Tanganyika, Lake Malawi, Lake Turkana, Lake Albert, Lake Kivu, and Lake Edward

3039. Why do some people think that the Great Lakes only consist of Lake Victoria, Lake Albert, and Lake Edward?
Because they are the only three that empty into the White Nile
3040. What are other names of Lake Victoria?
Victoria Nyanza, Ukerewe, Nalubaale, Sango, and Lolwe
3041. What is Lake Victoria's rank by size in Africa?
1st (26,600 sq mi / 68,800 km^2)
3042. What is Lake Victoria's rank by size in the world's freshwater lakes?
2nd
3043. What is the source of Lake Victoria?
The Kagera River
3044. Which countries are parts of the basin of Lake Victoria?
Tanzania, Uganda, and Kenya
3045. How many islands are there in Lake Victoria?
3,000 (Ssese Islands in Uganda)
3046. What is Lake Tanganyika's rank by volume among the world's freshwater lakes?
2nd (4,500 cu mi / 18,900 km^3)
3047. What is Lake Tanganyika's rank by depth in the world?
2nd (4,800 ft / 1,470 m)
3048. Which rivers flow into Lake Tanganyika?
Ruzizi River, Malagarasi River, and Kalambo River
3049. Which river flows out of Lake Tanganyika?
Lukuga River
3050. Lake Tanganyika is divided between which four countries?
Burundi, Democratic Republic of the Congo, Tanzania and Zambia
3051. What is the area of Lake Tanganyika?
12,700 sq mi / 32,900 km^2
3052. The Congo River (also known as the Zaire River) is the largest river in Western Central Africa. What is its rank by length in Africa?
2nd (2,922 mi / 4,700 km)
3053. The Congo River and its tributaries flow through the second largest rainforest area in the world. What is the rainforest name?
The Congo Rainforest
3054. What is the Congo River's rank by water flow in the world?
2nd (1,476,376 cu ft/s or 41,800 m^3/s)
3055. What is the Congo River's rank by drainage basin in the world?
2nd (1,420,848 sq mi / 3,680,000 km²)
3056. What is the Congo River's rank by depth in the world?
1st (>750 ft)
3057. Which countries does the Congo River pass through?
African Republic, Republic of the Congo, Angola, Zambia, Tanzania, Burundi, and Rwanda
3058. What is the Congo Basin?
The sedimentary basin that is the drainage of the Congo River of west equatorial Africa

3059. Congo is a traditional name for the region of equatorial Middle Africa that lies between the Gulf of Guinea and the African Great Lakes. What are countries wholly or partially in the Congo region?
Angola, Cameroon, Central African Republic, Democratic Republic of the Congo, Republic of the Congo, Equatorial Guinea, Gabon, and São Tomé and Príncipe
3060. What is the Zambezi River's rank by length in Africa?
4th (1,599 mi / 2,574 km)
3061. What is the third longest river in Africa?
The Niger River (2,611 mi / 4,200 km)
3062. Which countries does the Niger River run through?
Guinea, Mali, Niger, Benin, and Nigeria
3063. What is the main tributary of the Niger River?
The Benue River
3064. How long is the Benue River?
870 mi / 1,400 km
3065. The Benue River passes which countries?
Cameroon and Nigeria
3066. What is the mouth of the Niger River?
Gulf of Guinea in Nigeria
3067. What is the fourth longest river in Africa?
The Zambezi River
3068. Where is the origin of the Zambezi River?
Near Mwinilunga, Zambia
3069. Where is the mouth of the Zambezi River?
Indian Ocean
3070. Which countries does the Zambezi River pass through?
Botswana, Zimbabwe, Mozambique, Malawi, and Tanzania
3071. What is the Zambezi's most spectacular feature?
Victoria Falls
3072. Which two dams are main source of hydroelectric power on the Zambezi River?
The Kariba Dam (provides power to Zambia and Zimbabwe) and the Cahora Bassa Dam (provides power to both Mozambique and South Africa)
3073. What is the fifth longest river in Africa?
The Ubangi River (1,429 mi / 2,300 km)
3074. The Ubangi River is a tributary of which river?
The Congo River
3075. Which countries does the Ubangi River pass through?
Central African Republic, Democratic Republic of the Congo, and Republic of Congo
3076. What is the sixth longest river in Africa?
The Kasai River (1,338 mi / 2,153 km)
3077. The Kasai River is a tributary of which river?
The Congo River
3078. Which countries does the Ubangi River pass through?
Angola and the Democratic Republic of the Congo
3079. What is the seventh longest river in Africa?

The Orange River (1,300 mi /2,092 km)
3080. The Orange River is also called what other names?
Gariep River, Groote River or Senqu River
3081. Which countries does the Orange River pass through?
Lesotho, South Africa, and Namibia
3082. What is the source of the Orange River?
Senqu in Lesotho
3083. What is the mouth of the Orange River?
Alexander Bay in South Africa
3084. What is the eighth longest river in Africa?
The Limpopo River (1,118 mi / 1,800 km)
3085. Which countries does the Limpopo River pass through?
South Africa, Botswana, Zimbabwe, and Mozambique
3086. What is the mouth of the Limpopo River?
Indian Ocean
3087. What is the main tributary of the Limpopo River?
The Olifants River (Elephant River)
3088. The intersection of the Equator and Prime Meridian is located in which water body?
Gulf of Guinea
3089. What is the oceanic border of Gulf of Guinea?
The rhumb line that runs from Cape Palmas in Liberia to Cape Lopez in Gabon
3090. What body of water does the Niger River and the Volta River drain into?
Gulf of Guinea
3091. The coastline on Gulf of Guinea includes which two bights?
Bight of Benin and Bight of Bonny
3092. Which West African countries are located at the Bight of Bonny?
Nigeria, Cameroon, Equatorial Guinea (including Annobon), São Tomé and Príncipe and Gabon
3093. What is the East African mangroves?
An ecoregion consisting of mangrove swamps along the Indian Ocean coast of East Africa in southern Mozambique, Tanzania, Kenya and southern Somalia
3094. What are the Atlantic Equatorial coastal forests?
A tropical moist broadleaf forest ecoregion of central Africa, covering hills, plains, and mountains of the Atlantic coast of Cameroon, Equatorial Guinea, Gabon, Republic of the Congo, Angola, and Democratic Republic of the Congo
3095. What animals are found in the Atlantic Equatorial coastal forests?
Western Gorilla, chimpanzees, forest elephants and African buffalo, birds, amphibians, reptiles, invertebrates, black colobus monkeys and mandrills
3096. What are "Big Five" animals of Africa?
The five most difficult animals to be hunted on foot: lion, leopard, buffalo, rhinoceros and elephant
3097. What is Kaza Park (also called Kavango-Zambezi Transfrontier Conservation Area)?
The world's largest conservation area straddling Angola, Botswana, Namibia, Zambia and Zimbabwe (108,109 sq mi / 280,000 km^2)
3098. Kaza Park is scheduled to be completed by when?

2010
3099. What was Kingdom of Kongo?
An African kingdom located in west central Africa in present-day northern Angola, Cabinda, Republic of the Congo, and the western portion of Democratic Republic of the Congo
3100. When did Kingdom of Kongo exist?
1395–1914
3101. What was the capital of Kingdom of Kongo?
Mbanza-Kongo (Angola)
3102. What was the religion of Kingdom of Kongo?
Christianity with some traditional practices
3103. What kind of government did Kingdom of Kongo have?
Monarchy
3104. How large was Kingdom of Kongo?
49,962 sq mi / 129,400 km^2
3105. What was the population of Kingdom of Kongo?
509,250 (by 1650)
3106. What was the currency of Kingdom of Kongo?
Nzimbu Shells and Raffia Cloth
3107. What was the Berlin Conference (also called Congo Conference) during 1884–1885?
A conference regulated European colonization and trade in Africa during the New Imperialism period
3108. What was the Fashoda incident of 1898?
The climax of imperial territorial disputes between the United Kingdom and France in Eastern Africa, which brought the United Kingdom and France to the verge of war, but ended in a diplomatic victory for the UK
3109. The Scramble for Africa (also called the Race for Africa) was the result of conflicting European claims to African territory during which period?
Between the 1880s and the First World War in 1914
3110. What was German East Africa?
A German colony in East Africa, composed of present-day Burundi, Rwanda and Tanzania
3111. During which period did German East Africa exist?
February 27th, 1885–June 28th, 1919
3112. What was the capital of German East Africa during 1885-1890?
Bagamoyo (Tanzania)
3113. What was the capital of German East Africa during 1890-1919?
Dar es Salaam (Tanzania)
3114. How large was German East Africa in 1913?
384,172 sq mi (995,000 km^2)
3115. What was the population of German East Africa in 1913?
7,700,000
3116. What was the currency of German East Africa?
Rupie
3117. French West Africa was a federation of which eight French colonial territories?

Mauritania, Senegal, French Sudan (present-day Mali), French Guinea (present-day Guinea), Côte d'Ivoire, Upper Volta (present-day Burkina Faso), Dahomey (present-day Benin) and Niger

3118. During which period did French West Africa exist?
1895–1958

3119. What was the capital of French West Africa during 1895–1902?
Saint Louis

3120. What was the capital of French West Africa during 1902–1960?
Dakar

3121. What was the official language of French West Africa?
French

3122. What was the currency of French West Africa?
French West African franc

3123. French West Africa was preceded by which countries?
Senegambia and Niger, French Sudan, French Guinea, French Upper Volta, and French Dahomey

3124. French West Africa was succeeded by which countries?
French Community, Dahomey, Guinea, Ivory Coast, Mali Federation, Mauritania, Niger, and Upper Volta

3125. How large was French West Africa?
1,810,000 sq mi / 4,689,000 km²

3126. What was French Equatorial Africa?
The federation of French colonial possessions in Middle Africa

3127. Established in January 15th, 1910, French Equatorial Africa contained which four territories extending northwards from the Congo River to the Sahara Desert?
Gabon, Middle Congo (present-day Republic of the Congo), Oubangui-Chari (present-day Central African Republic) and Chad

3128. What was the capital of French Equatorial Africa?
Brazzaville

3129. What was the currency of French Equatorial Africa?
French Equatorial African Franc

3130. When was French Equatorial Africa disestablished?
September 1958

3131. What is Arab League?
A regional organization of Arab states in Southwest Asia, and North and Northeast Africa

3132. What is the official name of Arab League?
League of Arab States

3133. Where is headquarter of Arab League?
Cairo (Egypt)

3134. What is the area of Arab League?
5,382,910 sq mi / 13,953,041 km²

3135. What is the population of Arab League?
339,510,535 (by 2007)

3136. What is the official language of Arab League?
Arabic

3137. When was the Arab League formed?
    March 22nd, 1945
3138. Where was the Arab League formed?
    Cairo (Egypt)
3139. Who were the six original members of the Arab League?
    Egypt, Iraq, Transjordan (renamed Jordan after 1946), Lebanon, Saudi Arabia, and Syria
3140. Who are the 22 members of the Arab League now?
    Algeria, Bahrain, Comoros, Djibouti, Egypt, Iraq, Jordan, Kuwait, Lebanon, Libya, Mauritania, Morocco, Oman, Palestine, Qatar, Saudi Arabia, Somalia, Sudan, Syria, Tunisia, United Arab Emirates, and Yemen
3141. What is the African Union?
    An intergovernmental Organization consisting of 52 African countries
3142. Which country in Africa is not part of the African Union?
    Morocco
3143. What is the national anthem of the African Union?
    Let Us All Unite and Celebrate Together
3144. What are official languages of the African Union?
    Arabic, French, English, Portuguese, Swahili
3145. What kind of government does the African Union have?
    Continental union
3146. When was the African Union started as Organization of African Unity?
    May 25th, 1963
3147. When was the African Union established?
    July 9th, 2002
3148. What is the currency of the African Union?
    Afro (proposed for 2028)
3149. Where is the main administrative capital of the African Union, the place that the African Union Commission is headquartered?
    Addis Ababa (Ethiopia)

Bibliography

- Answers.com - Online Dictionary, Encyclopedia and much more, http://www.answers.com/
- A wilderness safari in Africa, http://www.zamtrade.com/
- Background Notes, http://www.state.gov/r/pa/ei/bgn/
- Canada's Aquatic Environments, http://www.aquatic.uoguelph.ca/
- CIA – The World Factbook, https://www.cia.gov/library/publications/the-world-factbook/
- Geography Page, http://peakbagger.com/
- Geography Summaries Index - Vaughn's Summaries, http://www.vaughns-1-pagers.com/geography/
- Infoplease: Encyclopedia, Almanac, Atlas, Biographies, Dictionary, Thesaurus. Free online reference, research & homework help, http://www.infoplease.com/
- Official global voting platform of New7Wonders, http://www.new7wonders.com/
- Perry-Castañeda Map Collection - UT Library Online, http://lib.utexas.edu/maps/
- Top 11 Largest Cities in the World in 2008 http://amolife.com/great-places/top-10-largest-cities-in-the-world.html
- Wikipedia – The Free Encyclopedia, http://en.wikipedia.org/wiki/Main_Page
- World Atlas of Maps Flags and Geography Facts and Figures, http://www.worldatlas.com/

Other Books

- World Geography Questionnaires: Americas - North America, Central America, South America and territories in the region (Volume 1), Kenneth Ma and Jennifer Fu, ISBN-10: 1449553222, ISBN-13: 978-1449553227
- The Missing Mau, Hermione Ma and Jennifer Fu, ISBN-10: 1451587090, ISBN-13: 978-1451587098

About the Authors

Kenneth Ma is an eleventh grader at Monta Vista High School in Cupertino, California. He was the National Geography School Bee at Eaton Elementary School in Cupertino when he was in fourth grade, and he was the National Geography School Bee at Kennedy Middle School in Cupertino when he was in sixth grade. He was also a member of the Kennedy Middle school National Geography Challenge Championship team in 2007. In 2007, he was honored with the Outstanding Achievement in Geography Award from his middle school. Besides his interest in Geography, Kenneth is also an avid soccer player.

Jennifer Fu is Kenneth's mom and lives in Cupertino, California. She is a software engineer by day and an aspiring writer by night. She has contributed short stories and novellas to a number of on-line publications in Chinese. Writing a geography book is a new endeavor for her.

CPSIA information can be obtained at www.ICGtesting.com
Printed in the USA
LVOW111917181211

259981LV00004B/119/P

9 781451 587074